STATISTICAL TABLES FOR THE SOCIAL, BIOLOGICAL AND PHYSICAL SCIENCES

compiled by

F. C. POWELL

T0254424

CAMBRIDGE UNIVERSITY PRESS

Cambridge
London   New York   New Rochelle
Melbourne   Sydney

CAMBRIDGE UNIVERSITY PRESS
Cambridge, New York, Melbourne, Madrid, Cape Town, Singapore, São Paulo, Delhi

Cambridge University Press
The Edinburgh Building, Cambridge CB2 8RU, UK

Published in the United States of America by Cambridge University Press, New York

www.cambridge.org
Information on this title: www.cambridge.org/9780521241410

First published 1982
Re-issued in this digitally printed version 2008

A catalogue record for this publication is available from the British Library

ISBN 978-0-521-24141-0 hardback
ISBN 978-0-521-28473-8 paperback

ACKNOWLEDGEMENTS

I am indebted to Dr P. M. E. Altham, of the Statistical Laboratory at the University of Cambridge, who read this compilation in manuscript and made many helpful comments and suggestions.

I wish to thank the Biometrika Trustees, Longman, the Rand Corporation and John Wiley for permission to use material from:

Biometrika

Pearson, E. S. & Hartley, H. O. (eds.) *Biometrika Tables for Statisticians* vols 1 and 2 (Cambridge University Press 1931, 1966)

Fisher, R. A. & Yates, F. *Statistical Tables for Biological, Agricultural and Medical Research* (Longman, London 1978)

The Rand Corporation. *A Million Random Digits* (The Free Press, Glencoe, Ill. 1955)

Conover, W. J. *Practical Nonparametric Statistics* (Wiley, New York 1971)

In addition, many of the tables are based on material which has appeared in the following:

Journals

Annals of Mathematical Statistics

Biometrics

Biometrika

Journal of the American Statistical Association

Books

Comrie, L. J. *Chambers's Shorter Six-Figure Mathematical Tables* (Chambers, Edinburgh 1966)

Harter, H. L. & Owen, D. B. *Selected Tables in Mathematical Statistics*, vol. 1, sponsored by the Institute of Mathematical Statistics (Markham, Chicago 1970)

Harvard University Computation Laboratory *Annals*, vol. 35: *Tables of the Cumulative Binomial Probability Distribution* (Harvard University Press 1955)

Hollander, M. & Wolfe, D. A. *Nonparametric Statistical Methods* (Wiley, New York 1973)

Miller, J. C. P. & Powell, F. C. *The Cambridge Elementary Mathematical Tables* (Cambridge University Press 1979)

Molina, E. C. *Poisson's Exponential Binomial Limit* (van Nostrand, New York 1942)

Owen, D. B. *Handbook of Statistical Tables* (Addison-Wesley, Reading, Mass. 1962)

F. C. POWELL

CONTENTS

EXPLANATION OF STATISTICAL TERMS AND PROCEDURES

Tail probabilities. The *lower tail probability* P and the *upper tail probability* Q of a random variable X are defined as:
$$P(x) = \text{Prob}\,(X \leqslant x) \qquad Q(x) = \text{Prob}\,(X \geqslant x)$$

$P(x)$ is also called the *cumulative probability distribution function*. If $P(x)$ and $Q(x)$ are continuous functions of x:
$$P(x) + Q(x) = 1$$

If X is a discrete variable taking integer values:
$$P(x) + Q(x+1) = 1$$

Quantiles. There is no generally agreed definition or notation for these. Here the following definitions are adopted. For $P \leqslant \frac{1}{2}$ the *lower quantile* $x_{[P]}$ is the smallest x such that $\text{Prob}\,(X \leqslant x) \geqslant P$. Equivalently:
$$\text{Prob}\,(X \leqslant x_{[P]}) \geqslant P \qquad \text{Prob}\,(X \leqslant x) < P \quad \text{if} \quad x < x_{[P]}$$

For $P \geqslant \frac{1}{2}$ the *upper quantile* $x_{[P]}$ is the largest x such that $\text{Prob}\,(X \geqslant x) \geqslant 1 - P$ or Q. Equivalently:
$$\text{Prob}\,(X \geqslant x_{[P]}) \geqslant 1 - P \qquad \text{Prob}\,(X \geqslant x) < 1 - P \quad \text{if} \quad x > x_{[P]}$$

For a continuous distribution (more precisely a distribution having a continuous and strictly increasing cumulative distribution function) these definitions reduce to:
$$\text{Prob}\,(X < x_{[P]}) = P \quad \text{or equivalently} \quad \text{Prob}\,(X > x_{[P]}) = 1 - P \text{ or } Q$$

Graphical illustrations. (1) Consider first a variable X having the standard normal probability distribution. In fig. 1 the ordinate represents the probability density; the areas under the graph represent tail probabilities. The functions $P(x)$ and $Q(x)$ are graphed in fig. 2. In fig. 3 $x_{[P]}$ is graphed as a function of P.

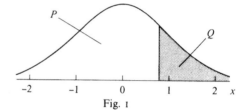

Fig. 1

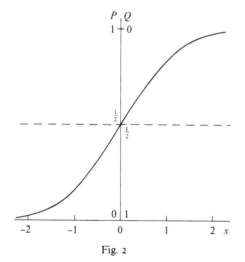

Fig. 2

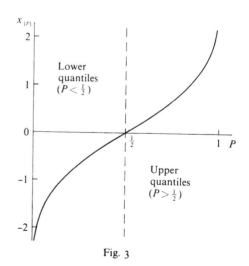

Fig. 3

6

(2) Next consider a variable X having the binomial distribution $\text{Bi}(2, \frac{1}{3})$.† It takes values 0, 1, 2 with probabilities $\frac{4}{9}, \frac{4}{9}, \frac{1}{9}$ respectively. The tail probabilities are now step functions of x, as shown in fig. 4; the values of $P(x)$ and $Q(x)$ for $x = 0$, 1, 2 are indicated by crosses and circles respectively.

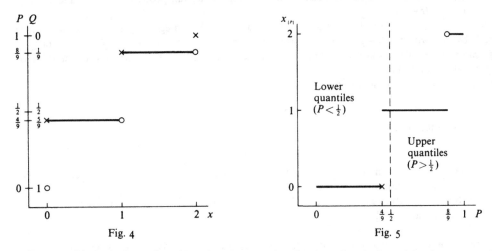

Fig. 4   Fig. 5

In fig. 5 $x_{[P]}$ is graphed as a function of P. For $P = \frac{4}{9}$ and $\frac{8}{9}$ the values indicated by the cross and the circle are to be taken.

Interpolation. Many of the tables give probabilities or quantiles for a family of distributions specified by the values of one or more parameters, e.g. $t(\nu)$, $\chi^2(\nu)$, $\text{Bi}(n, \pi)$. It is not usually necessary or possible to include all values of the parameters and use must be made of interpolation. These tables have been constructed with *linear interpolation* in mind; tabular intervals and the number of significant figures have been chosen accordingly.

Linear interpolation is based on the assumption that the tabulated function changes at a constant rate between one tabulated value and the next. If successive tabular entries y_1 and $y_2 = y_1 + \Delta$ correspond to values x_1 and $x_2 = x_1 + h$ of the independent variable, the value of the function corresponding to an intermediate value $x = x_1 + \theta h$ (where $0 < \theta < 1$) is taken to be $y = y_1 + \theta \Delta$ or, equivalently, $(1 - \theta)y_1 + \theta y_2$. For example, suppose we have to find the upper tail probability $Q(1)$ for the binomial distribution $\text{Bi}(2, \pi)$ with $\pi = 0.408$. Now for $\pi = 0.40$ the value of $Q(1)$ is 0.6400, for $\pi = 0.42$ it is 0.6636 (see p. 15), and so $\Delta = 0.0236$. For $\pi = 0.408$, $\theta = 0.4$, and so $Q(1) \approx 0.6400 + 0.4 \times 0.0236 \approx 0.6494$. Alternatively we may calculate $Q(1)$ as $0.6 \times 0.6400 + 0.4 \times 0.6636$ with the same result.

The errors introduced by linear interpolation should be borne in mind. Where tabulation is at equal intervals of a parameter, these errors cannot exceed $1 + S/8$ units in the last decimal place, where S is the absolute value of the second difference of the tabular values (i.e. the difference between successive values of the first difference Δ). In the example above, $S = 8$ and so the error cannot exceed 2 units. The true value is in fact 0.649536 and the error is 0.000136.

In a number of tables interpolation should be linear in an inverse power of the parameter (usually n or ν). Suppose, for example, we wish to find $t_{[.999]}(50)$. From the table on p. 39 we have $t_{[.999]}(40) = 3.307$ and $t_{[.999]}(60) = 3.232$; interpolation should be linear in $120/\nu$. Now $120/40 = 3$, $120/60 = 2$ and $120/50 = 2.4$, so

$$t_{[.999]}(50) \approx 3.307 + \frac{2.4 - 3}{2 - 3} \times (3.232 - 3.307)$$

$$= 3.307 - 0.6 \times 0.075$$

$$= 3.262$$

The true value is $3.261\ldots$

---

† We write $X \sim \text{Bi}(2, \frac{1}{3})$.

Approximations. It may be that the probability distribution of a discrete random variable r approximates to the known distribution of a continuous random variable, e.g. $N(\mu, \sigma^2)$. Then $N(0, 1)$ is an approximation to the distribution of $(r-\mu)/\sigma$. This means that the tail probabilities $P(r)$ and $Q(r)$ of the r-distribution are nearly equal to the corresponding tail probabilities $P(z)$ and $Q(z)$ of $N(0, 1)$ with $z = (r-\mu)/\sigma$.

A *correction for continuity* should however be made. If r takes integer values, $P(r)$ is given more correctly by $P(z)$ with $z = (r-\mu+\frac{1}{2})/\sigma$ and $Q(r)$ is given by $Q(z)$ with $z = (r-\mu-\frac{1}{2})/\sigma$. Values of $Q(z)$ are given on p. 31. We can also use $z = (r-\mu-\frac{1}{2})/\sigma$ as a test statistic by referring it to the table of upper quantiles on p. 30.

Lower quantiles of r are given approximately by:

$$r_{[P]} \approx \mu + \sigma z_{[P]} - \tfrac{1}{2}$$

and upper quantiles by:

$$r_{[P]} \approx \mu + \sigma z_{[P]} + \tfrac{1}{2}$$

These corrections for continuity are not usually recommended if the approximation is conservative (i.e. if it overestimates tail probabilities).

In some tables the variable r takes even values only or odd values only. Corrections of ± 1 should then be made instead of $\pm\frac{1}{2}$.

Tests of hypotheses. The procedure and terminology used in many tests can be explained by means of an example. Suppose that a theory predicts that the value x of a physical quantity is 10 units, the alternative being $x < 10$. To test the theory, a series of twelve measurements is made by a method that is liable to a normally distributed error with zero mean and standard deviation 6 units. The average of the twelve measurements is 6.3 units. Does this result conflict with the theory?

The twelve measurements can be regarded as a *random sample* from an *infinite population* of measurements having a normal probability distribution with unknown mean μ and variance 6^2, $N(\mu, 6^2)$. The object is to decide between the *null hypothesis* H_0 that $\mu = 10$ and the *alternative hypothesis* H_1 that $\mu < 10$. As *test statistic* we take the sample mean $\bar{x}$. If we repeatedly take samples of size twelve from the population we build up the *sampling distribution* of $\bar{x}$; this is a normal distribution with mean μ and variance $6^2/12$ (see p. 46). Under H_0 the sampling distribution of $\bar{x}$ is thus $N(10, 3)$; this is the *null distribution* of $\bar{x}$.

If m is a value of $\bar{x}$ in the extreme lower tail of the null distribution, Prob $(\bar{x} \leqslant m)$ is smaller for the null distribution than for a distribution with $\mu < 10$ (see fig. 6). If the observed value of $\bar{x}$ is less

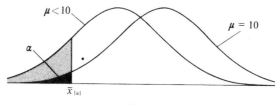

Fig. 6

than m it is plausible to say that $\mu < 10$ is more likely than $\mu = 10$. Accordingly we choose a *significance level* α (conventionally often 0.01 or 0.05, i.e. 1% or 5%) and, taking m to be the quantile $\bar{x}_{[\alpha]}$, we declare that *at significance level α we can reject the null hypothesis H_0 in favour of the alternative hypothesis H_1 if the observed value of the test statistic lies in the rejection (or critical) region* $\bar{x} < \bar{x}_{[\alpha]}$.

Since the null distribution of $\bar{x}$ is $N(10, 3)$, $\bar{x}_{[\alpha]}$ for this distribution is $10 + \sqrt{3}\, z_{[\alpha]}$, where $z_{[\alpha]}$ is a quantile of $N(0, 1)$. Taking values of $z_{[\alpha]}$ from the table on p. 30, we have:

$$\bar{x}_{[.01]} = 10 - \sqrt{3} \times 2.326 = 5.97 \qquad \bar{x}_{[.05]} = 10 - \sqrt{3} \times 1.645 = 7.15$$

The observed value of $\bar{x}$, 6.3, lies between these quantiles. Accordingly we reject the null hypothesis (that $\mu = 10$) in favour of the alternative hypothesis (that $\mu < 10$) at significance level 0.05 but not at 0.01.

One-sided and two-sided tests. In the example discussed above the alternative hypothesis is that $\mu < 10$; the test is thus *one-sided*. The rejection region of the null distribution at significance level α is in the lower tail of the null distribution; thus the test is also a *one-tail* test.

If, instead, the alternative hypothesis is taken to be $\mu > 10$, the test is again one-sided. The appropriate rejection region at significance level α is defined by $\bar{x} > \bar{x}_{[1-\alpha]}$, and again the test is a one-tail test.

If the alternative hypothesis is taken to be $\mu \neq 10$ we have a *two-sided* test. We now reject H_0 in favour of H_1 at significance level α if $\bar{x} < \bar{x}_{[\alpha/2]}$ or $\bar{x} > \bar{x}_{[1-\alpha/2]}$. The rejection region lies partly in the lower tail and partly in the upper tail of the null distribution; thus we have a *two-tail* test.

Two-sided tests are not necessarily two-tailed. For instance, if a random sample has been taken from a normal population with unknown mean and variance, the null hypothesis $\mu = \mu_0$ can be tested against the alternative hypothesis $\mu \neq \mu_0$ either by a two-tail t-test or by a one-tail F-test; see p. 46.

Critical values. Consider first a test statistic U taking *discrete values*. For a rejection region $U < u_{[\alpha]}$, the *critical value* $u_{(\alpha)}$ is defined as the largest value of U in the region; it is the next value of U below $u_{[\alpha]}$. If U takes integer values:

$$u_{(\alpha)} = u_{[\alpha]} - 1$$

The rejection region is $U \leqslant u_{(\alpha)}$.

For a rejection region $U > u_{[1-\alpha]}$, the *critical value* $u_{(1-\alpha)}$ is the smallest value of u in the region; it is the next value of U above $u_{[1-\alpha]}$. If U takes integer values:

$$u_{(1-\alpha)} = u_{[1-\alpha]} + 1$$

The rejection region is $U \geqslant u_{(1-\alpha)}$.

Note that a rejection region contains the corresponding critical value but not the corresponding quantile.

For a test statistic U with a *continuous range of values*, critical values are identical with the corresponding quantiles:

$$u_{(\alpha)} = u_{[\alpha]} \qquad u_{(1-\alpha)} = u_{[1-\alpha]}$$

Ties. In some tests involving the ranking of observations, according to their magnitudes perhaps, special procedures are required for assessing significance when ties occur; these are not described here. It is however often sufficient to assign to each of a tied group of observations their 'average rank', i.e. the average of the ranks they would have been given if their magnitudes had been unequal. If, for instance, four observations are tied 'equal fifth', they are each given the rank $(5+6+7+8)/4 = 6\frac{1}{2}$. In some tests tied observations can safely be discarded provided that they are not too numerous.

Confidence intervals. Consider again the example discussed above. Suppose that the average of twelve measurements of a physical quantity, by a method liable to a normally distributed error with zero mean and standard deviation 6 units, is 6.3 units. If the true value is μ, the probability distribution of the measured value x is $N(\mu, 36)$ and the distribution of the sample mean $\bar{x}$ is $N(\mu, 3)$. Therefore, with probability $1-\alpha$,

$$\bar{x} - \mu > \sqrt{3}\, z_{[\alpha]}$$

where $z_{[\alpha]}$ is a quantile of $N(0, 1)$, so that

$$\mu < \bar{x} - \sqrt{3}\, z_{[\alpha]}$$

For given $\bar{x}$ this defines a *one-sided confidence interval* for μ. At confidence level $1-\alpha$ we may expect μ to lie in the interval (but note that we cannot say that μ lies in the interval with probability $1-\alpha$).

For $\alpha = 0.01$ and $\bar{x} = 6.3$, the upper bound of the confidence interval is $6.3 + \sqrt{3} \times 2.326$ or 10.33; for $\alpha = 0.05$ it is $6.3 + \sqrt{3} \times 1.645$ or 9.15. Therefore the predicted value of 10 units does not conflict with the measurements at confidence level 0.99, but it does conflict at level 0.95. This conclusion agrees with that reached earlier by a different procedure.

Upper and lower bounds of *two-sided confidence intervals* can be found in a similar way. An example is given on p. 47.

FACTORIALS AND LOGARITHMS OF FACTORIALS (1–199)

n	$\dagger n!$	lg $n!$	n	$\dagger n!$	lg $n!$	n	$\dagger n!$	lg $n!$	n	$\dagger n!$	lg $n!$
0	1	0.0000	50	3.0414	64.4831	100	9.3326	157.9700	150	5.7134	262.7569
1	1	0.0000	51	1.5511	66.1906	101	9.4259	159.9743	151	8.6272	264.9359
2	2	0.3010	52	8.0658	67.9066	102	9.6145	161.9829	152	1.3113	267.1177
3	6	0.7782	53	4.2749	69.6309	103	9.9029	163.9958	153	2.0063	269.3024
4	24	1.3802	54	2.3084	71.3633	104	1.0299	166.0128	154	3.0898	271.4899
5	120	2.0792	55	1.2696	73.1037	105	1.0814	168.0340	155	4.7891	273.6803
6	720	2.8573	56	7.1100	74.8519	106	1.1463	170.0593	156	7.4711	275.8734
7	5040	3.7024	57	4.0527	76.6077	107	1.2265	172.0887	157	1.1730	278.0693
8	40320	4.6055	58	2.3506	78.3712	108	1.3246	174.1221	158	1.8533	280.2679
9	3.6288	5.5598	59	1.3868	80.1420	109	1.4439	176.1595	159	2.9467	282.4693
10	3.6288	6.5598	60	8.3210	81.9202	110	1.5882	178.2009	160	4.7147	284.6735
11	3.9917	7.6012	61	5.0758	83.7055	111	1.7630	180.2462	161	7.5907	286.8803
12	4.7900	8.6803	62	3.1470	85.4979	112	1.9745	182.2955	162	1.2297	289.0898
13	6.2270	9.7943	63	1.9826	87.2972	113	2.2312	184.3485	163	2.0044	291.3020
14	8.7178	10.9404	64	1.2689	89.1034	114	2.5436	186.4054	164	3.2872	293.5168
15	1.3077	12.1165	65	8.2477	90.9163	115	2.9251	188.4661	165	5.4239	295.7343
16	2.0923	13.3206	66	5.4434	92.7359	116	3.3931	190.5306	166	9.0037	297.9544
17	3.5569	14.5511	67	3.6471	94.5619	117	3.9699	192.5988	167	1.5036	300.1771
18	6.4024	15.8063	68	2.4800	96.3945	118	4.6845	194.6707	168	2.5261	302.4024
19	1.2165	17.0851	69	1.7112	98.2333	119	5.5746	196.7462	169	4.2691	304.6303
20	2.4329	18.3861	70	1.1979	100.0784	120	6.6895	198.8254	170	7.2574	306.8608
21	5.1091	19.7083	71	8.5048	101.9297	121	8.0943	200.9082	171	1.2410	309.0938
22	1.1240	21.0508	72	6.1234	103.7870	122	9.8750	202.9945	172	2.1346	311.3293
23	2.5852	22.4125	73	4.4701	105.6503	123	1.2146	205.0844	173	3.6928	313.5674
24	6.2045	23.7927	74	3.3079	107.5196	124	1.5061	207.1779	174	6.4254	315.8079
25	1.5511	25.1906	75	2.4809	109.3946	125	1.8827	209.2748	175	1.1244	318.0509
26	4.0329	26.6056	76	1.8855	111.2754	126	2.3722	211.3751	176	1.9790	320.2965
27	1.0889	28.0370	77	1.4518	113.1619	127	3.0127	213.4790	177	3.5029	322.5444
28	3.0489	29.4841	78	1.1324	115.0540	128	3.8562	215.5862	178	6.2351	324.7948
29	8.8418	30.9465	79	8.9462	116.9516	129	4.9745	217.6967	179	1.1161	327.0477
30	2.6525	32.4237	80	7.1569	118.8547	130	6.4669	219.8107	180	2.0090	329.3030
31	8.2228	33.9150	81	5.7971	120.7632	131	8.4716	221.9280	181	3.6362	331.5606
32	2.6313	35.4202	82	4.7536	122.6770	132	1.1182	224.0485	182	6.6179	333.8207
33	8.6833	36.9387	83	3.9455	124.5961	133	1.4873	226.1724	183	1.2111	336.0832
34	2.9523	38.4702	84	3.3142	126.5204	134	1.9929	228.2995	184	2.2284	338.3480
35	1.0333	40.0142	85	2.8171	128.4498	135	2.6905	230.4298	185	4.1225	340.6152
36	3.7199	41.5705	86	2.4227	130.3843	136	3.6590	232.5634	186	7.6679	342.8847
37	1.3764	43.1387	87	2.1078	132.3238	137	5.0129	234.7001	187	1.4339	345.1565
38	5.2302	44.7185	88	1.8548	134.2683	138	6.9178	236.8400	188	2.6957	347.4307
39	2.0398	46.3096	89	1.6508	136.2177	139	9.6157	238.9830	189	5.0949	349.7071
40	8.1592	47.9116	90	1.4857	138.1719	140	1.3462	241.1291	190	9.6803	351.9859
41	3.3453	49.5244	91	1.3520	140.1310	141	1.8981	243.2783	191	1.8489	354.2669
42	1.4050	51.1477	92	1.2438	142.0948	142	2.6954	245.4306	192	3.5500	356.5502
43	6.0415	52.7811	93	1.1568	144.0632	143	3.8544	247.5860	193	6.8514	358.8358
44	2.6583	54.4246	94	1.0874	146.0364	144	5.5503	249.7443	194	1.3292	361.1236
45	1.1962	56.0778	95	1.0330	148.0141	145	8.0479	251.9057	195	2.5919	363.4136
46	5.5026	57.7406	96	9.9168	149.9964	146	1.1750	254.0700	196	5.0801	365.7059
47	2.5862	59.4127	97	9.6193	151.9831	147	1.7272	256.2374	197	1.0008	368.0003
48	1.2414	61.0939	98	9.4269	153.9744	148	2.5563	258.4076	198	1.9816	370.2970
49	6.0828	62.7841	99	9.3326	155.9700	149	3.8089	260.5808	199	3.9433	372.5959

† If n is greater than 8, multiply the tabulated value by 10^c, where c is the integer part of lg $n!$. Thus $99! = 9.3326 \times 10^{155}$.

The definition of $n!$ is:

$$n! = n(n-1)(n-2)\ldots 1 \text{ for positive integers } n$$
$$0! = 1$$

n	$\dagger n!$	$\lg n!$	n	$\dagger n!$	$\lg n!$	n	$\dagger n!$	$\lg n!$	n	$\dagger n!$	$\lg n!$
200	7.8866	374.8969	250	3.2329	492.5096	300	3.0606	614.4858	350	1.2359	740.0920
201	1.5852	377.2001	251	8.1145	494.9093	301	9.2123	616.9644	351	4.3379	742.6373
202	3.2021	379.5054	252	2.0448	497.3107	302	2.7821	619.4444	352	1.5269	745.1838
203	6.5003	381.8129	253	5.1735	499.7138	303	8.4298	621.9258	353	5.3901	747.7316
204	1.3261	384.1226	254	1.3141	502.1186	304	2.5627	624.4087	354	1.9081	750.2806
205	2.7184	386.4343	255	3.3509	504.5252	305	7.8161	626.8930	355	6.7738	752.8308
206	5.5999	388.7482	256	8.5782	506.9334	306	2.3917	629.3787	356	2.4115	755.3823
207	1.1592	391.0642	257	2.2046	509.3433	307	7.3426	631.8659	357	8.6089	757.9349
208	2.4111	393.3822	258	5.6878	511.7549	308	2.2615	634.3544	358	3.0820	760.4888
209	5.0392	395.7024	259	1.4732	514.1682	309	6.9881	636.8444	359	1.1064	763.0439
210	1.0582	398.0246	260	3.8302	516.5832	310	2.1663	639.3357	360	3.9832	765.6002
211	2.2329	400.3489	261	9.9968	518.9999	311	6.7373	641.8285	361	1.4379	768.1577
212	4.7337	402.6752	262	2.6192	521.4182	312	2.1020	644.3226	362	5.2053	770.7164
213	1.0083	405.0036	263	6.8884	523.8381	313	6.5793	646.8182	363	1.8895	773.2764
214	2.1577	407.3340	264	1.8185	526.2597	314	2.0659	649.3151	364	6.8778	775.8375
215	4.6391	409.6664	265	4.8191	528.6830	315	6.5076	651.8134	365	2.5104	778.3997
216	1.0020	412.0009	266	1.2819	531.1078	316	2.0564	654.3131	366	9.1881	780.9632
217	2.1744	414.3373	267	3.4226	533.5344	317	6.5188	656.8142	367	3.3720	783.5279
218	4.7403	416.6758	268	9.1727	535.9625	318	2.0730	659.3166	368	1.2409	786.0937
219	1.0381	419.0162	269	2.4674	538.3922	319	6.6128	661.8204	369	4.5790	788.6608
220	2.2839	421.3587	270	6.6621	540.8236	320	2.1161	664.3255	370	1.6942	791.2290
221	5.0473	423.7031	271	1.8054	543.2566	321	6.7927	666.8320	371	6.2855	793.7983
222	1.1205	426.0494	272	4.9108	545.6912	322	2.1872	669.3399	372	2.3382	796.3689
223	2.4987	428.3977	273	1.3406	548.1273	323	7.0648	671.8491	373	8.7216	798.9406
224	5.5972	430.7480	274	3.6734	550.5651	324	2.2890	674.3596	374	3.2619	801.5135
225	1.2594	433.1002	275	1.0102	553.0044	325	7.4392	676.8715	375	1.2232	804.0875
226	2.8462	435.4543	276	2.7881	555.4453	326	2.4252	679.3847	376	4.5992	806.6627
227	6.4608	437.8103	277	7.7230	557.8878	327	7.9304	681.8993	377	1.7339	809.2390
228	1.4731	440.1682	278	2.1470	560.3318	328	2.6012	684.4152	378	6.5542	811.8165
229	3.3733	442.5281	279	5.9901	562.7774	329	8.5578	686.9324	379	2.4840	814.3952
230	7.7586	444.8898	280	1.6772	565.2246	330	2.8241	689.4509	380	9.4393	816.9749
231	1.7922	447.2534	281	4.7130	567.6733	331	9.3477	691.9707	381	3.5964	819.5559
232	4.1580	449.6189	282	1.3291	570.1235	332	3.1034	694.4918	382	1.3738	822.1379
233	9.6881	451.9862	283	3.7613	572.5753	333	1.0334	697.0143	383	5.2617	824.7211
234	2.2670	454.3555	284	1.0682	575.0287	334	3.4517	699.5380	384	2.0205	827.3055
235	5.3275	456.7265	285	3.0444	577.4835	335	1.1563	702.0631	385	7.7789	829.8909
236	1.2573	459.0994	286	8.7069	579.9399	336	3.8852	704.5894	386	3.0027	832.4775
237	2.9798	461.4742	287	2.4989	582.3977	337	1.3093	707.1170	387	1.1620	835.0652
238	7.0918	463.8508	288	7.1968	584.8571	338	4.4255	709.6460	388	4.5087	837.6540
239	1.6950	466.2292	289	2.0799	587.3180	339	1.5003	712.1762	389	1.7539	840.2440
240	4.0679	468.6094	290	6.0316	589.7804	340	5.1009	714.7076	390	6.8401	842.8351
241	9.8036	470.9914	291	1.7552	592.2443	341	1.7394	717.2404	391	2.6745	845.4272
242	2.3725	473.3752	292	5.1252	594.7097	342	5.9487	719.7744	392	1.0484	848.0205
243	5.7651	475.7608	293	1.5017	597.1766	343	2.0404	722.3097	393	4.1202	850.6149
244	1.4067	478.1482	294	4.4149	599.6449	344	7.0190	724.8463	394	1.6234	853.2104
245	3.4464	480.5374	295	1.3024	602.1147	345	2.4216	727.3841	395	6.4123	855.8070
246	8.4781	482.9283	296	3.8551	604.5860	346	8.3786	729.9232	396	2.5393	858.4047
247	2.0941	485.3210	297	1.1450	607.0588	347	2.9074	732.4635	397	1.0081	861.0035
248	5.1933	487.7155	298	3.4120	609.5330	348	1.0118	735.0051	398	4.0122	863.6034
249	1.2931	490.1116	299	1.0202	612.0087	349	3.5311	737.5479	399	1.6009	866.2044

† Multiply the tabulated value by 10^c, where c is the integer part of $\lg n!$. Thus $390! = 6.8401 \times 10^{842}$.

Source: For the tables on pp. 10–12, *Chambers's Shorter Six-Figure Mathematical Tables* (L. J. Comrie).

FACTORIALS AND LOGARITHMS OF FACTORIALS (400–600)

n	$\dagger n!$	$\lg n!$	n	$\dagger n!$	$\lg n!$	n	$\dagger n!$	$\lg n!$	n	$\dagger n!$	$\lg n!$
400	6.4035	868.8064	450	1.7334	1000.2389	500	1.2201	1134.0864	550	1.2789	1270.1069
401	2.5678	871.4096	451	7.8175	1002.8931	501	6.1129	1136.7862	551	7.0470	1272.8480
402	1.0322	874.0138	452	3.5335	1005.5482	502	3.0687	1139.4870	552	3.8899	1275.5899
403	4.1600	876.6191	453	1.6007	1008.2043	503	1.5435	1142.1885	553	2.1511	1278.3327
404	1.6806	879.2255	454	7.2671	1010.8614	504	7.7794	1144.8909	554	1.1917	1281.0762
405	6.8065	881.8329	455	3.3065	1013.5194	505	3.9286	1147.5942	555	6.6141	1283.8205
406	2.7635	884.4415	456	1.5078	1016.1783	506	1.9879	1150.2984	556	3.6774	1286.5655
407	1.1247	887.0510	457	6.8905	1018.8383	507	1.0079	1153.0034	557	2.0483	1289.3114
408	4.5889	889.6617	458	3.1559	1021.4991	508	5.1199	1155.7093	558	1.1430	1292.0580
409	1.8769	892.2734	459	1.4485	1024.1609	509	2.6060	1158.4160	559	6.3892	1294.8054
410	7.6951	894.8862	460	6.6633	1026.8237	510	1.3291	1161.1236	560	3.5779	1297.5536
411	3.1627	897.5001	461	3.0718	1029.4874	511	6.7916	1163.8320	561	2.0072	1300.3026
412	1.3030	900.1150	462	1.4192	1032.1520	512	3.4773	1166.5412	562	1.1281	1303.0523
413	5.3815	902.7309	463	6.5707	1034.8176	513	1.7838	1169.2514	563	6.3510	1305.8028
414	2.2279	905.3479	464	3.0488	1037.4841	514	9.1690	1171.9623	564	3.5820	1308.5541
415	9.2459	907.9660	465	1.4177	1040.1516	515	4.7220	1174.6741	565	2.0238	1311.3062
416	3.8463	910.5850	466	6.6065	1042.8200	516	2.4366	1177.3868	566	1.1455	1314.0590
417	1.6039	913.2052	467	3.0852	1045.4893	517	1.2597	1180.1003	567	6.4948	1316.8126
418	6.7044	915.8264	468	1.4439	1048.1595	518	6.5253	1182.8146	568	3.6891	1319.5669
419	2.8091	918.4486	469	6.7718	1050.8307	519	3.3866	1185.5298	569	2.0991	1322.3220
420	1.1798	921.0718	470	3.1828	1053.5028	520	1.7610	1188.2458	570	1.1965	1325.0779
421	4.9671	923.6961	471	1.4991	1056.1758	521	9.1750	1190.9626	571	6.8319	1327.8345
422	2.0961	926.3214	472	7.0756	1058.8498	522	4.7894	1193.6803	572	3.9078	1330.5919
423	8.8666	928.9478	473	3.3468	1061.5246	523	2.5048	1196.3988	573	2.2392	1333.3501
424	3.7594	931.5751	474	1.5864	1064.2004	524	1.3125	1199.1181	574	1.2853	1336.1090
425	1.5978	934.2035	475	7.5353	1066.8771	525	6.8908	1201.8383	575	7.3904	1338.8687
426	6.8064	936.8329	476	3.5868	1069.5547	526	3.6246	1204.5593	576	4.2569	1341.6291
427	2.9063	939.4633	477	1.7109	1072.2332	527	1.9101	1207.2811	577	2.4562	1344.3903
428	1.2439	942.0948	478	8.1781	1074.9127	528	1.0086	1210.0037	578	1.4197	1347.1522
429	5.3364	944.7272	479	3.9173	1077.5930	529	5.3353	1212.7272	579	8.2201	1349.9149
430	2.2947	947.3607	480	1.8803	1080.2742	530	2.8277	1215.4514	580	4.7676	1352.6783
431	9.8900	949.9952	481	9.0443	1082.9564	531	1.5015	1218.1765	581	2.7700	1355.4425
432	4.2725	952.6307	482	4.3593	1085.6394	532	7.9880	1220.9024	582	1.6121	1358.2074
433	1.8500	955.2672	483	2.1056	1088.3234	533	4.2576	1223.6292	583	9.3988	1360.9731
434	8.0289	957.9047	484	1.0191	1091.0082	534	2.2736	1226.3567	584	5.4889	1363.7395
435	3.4926	960.5431	485	4.9426	1093.6940	535	1.2164	1229.0851	585	3.2110	1366.5066
436	1.5228	963.1826	486	2.4021	1096.3806	536	6.5197	1231.8142	586	1.8816	1369.2745
437	6.6545	965.8231	487	1.1698	1099.0681	537	3.5011	1234.5442	587	1.1045	1372.0432
438	2.9147	968.4646	488	5.7087	1101.7565	538	1.8836	1237.2750	588	6.4946	1374.8126
439	1.2795	971.1071	489	2.7916	1104.4458	539	1.0152	1240.0066	589	3.8253	1377.5827
440	5.6299	973.7505	490	1.3679	1107.1360	540	5.4823	1242.7390	590	2.2569	1380.3535
441	2.4828	976.3949	491	6.7162	1109.8271	541	2.9659	1245.4722	591	1.3339	1383.1251
442	1.0974	979.0404	492	3.3044	1112.5191	542	1.6075	1248.2062	592	7.8964	1385.8974
443	4.8615	981.6868	493	1.6291	1115.2119	543	8.7289	1250.9410	593	4.6826	1388.6705
444	2.1585	984.3342	494	8.0476	1117.9057	544	4.7485	1253.6766	594	2.7814	1391.4443
445	9.6053	986.9825	495	3.9836	1120.6003	545	2.5879	1256.4130	595	1.6550	1394.2188
446	4.2840	989.6318	496	1.9758	1123.2958	546	1.4130	1259.1501	596	9.8636	1396.9940
447	1.9149	992.2822	497	9.8199	1125.9921	547	7.7292	1261.8881	597	5.8885	1399.7700
448	8.5789	994.9334	498	4.8903	1128.6893	548	4.2356	1264.6269	598	3.5213	1402.5467
449	3.8519	997.5857	499	2.4403	1131.3874	549	2.3254	1267.3665	599	2.1093	1405.3241
450	1.7334	1000.2389	500	1.2201	1134.0864	550	1.2789	1270.1069	600	1.2656	1408.1023

$\dagger$ Multiply the tabulated value by 10^c, where c is the integer part of $\lg n!$. Thus $450! = 1.7334 \times 10^{1000}$.

For large n,
$$\ln n! = \ln \sqrt{(2\pi)} + (n + \tfrac{1}{2})\ln n - \left(n - \frac{1}{12n} + \dots\right)$$

which leads to:
$$\lg n! \approx 0.3991 + (n + \tfrac{1}{2})\lg n - 0.4342945\,n + 0.036/n$$

BINOMIAL COEFFICIENTS

n	$\binom{n}{0}$	$\binom{n}{1}$	$\binom{n}{2}$	$\binom{n}{3}$	$\binom{n}{4}$	$\binom{n}{5}$	$\binom{n}{6}$	$\binom{n}{7}$	$\binom{n}{8}$	$\binom{n}{9}$	$\binom{n}{10}$	$\binom{n}{11}$	$\binom{n}{12}$
1	1	1											
2	1	2	1										
3	1	3	3	1									
4	1	4	6	4	1								
5	1	5	10	10	5	1							
6	1	6	15	20	15	6	1						
7	1	7	21	35	35	21	7	1					
8	1	8	28	56	70	56	28	8	1				
9	1	9	36	84	126	126	84	36	9	1			
10	1	10	45	120	210	252	210	120	45	10	1		
11	1	11	55	165	330	462	462	330	165	55	11	1	
12	1	12	66	220	495	792	924	792	495	220	66	12	1
13	1	13	78	286	715	1287	1716	1716	1287	715	286	78	13
14	1	14	91	364	1001	2002	3003	3432	3003	2002	1001	364	91
15	1	15	105	455	1365	3003	5005	6435	6435	5005	3003	1365	455
16	1	16	120	560	1820	4368	8008	11440	12870	11440	8008	4368	1820
17	1	17	136	680	2380	6188	12376	19448	24310	24310	19448	12376	6188
18	1	18	153	816	3060	8568	18564	31824	43758	48620	43758	31824	18564
19	1	19	171	969	3876	11628	27132	50388	75582	92378	92378	75582	50388
20	1	20	190	1140	4845	15504	38760	77520	125970	167960	184756	167960	125970
21	1	21	210	1330	5985	20349	54264	116280	203490	293930	352716	352716	293930
22	1	22	231	1540	7315	26334	74613	170544	319770	497420	646646	705432	646646
23	1	23	253	1771	8855	33649	100947	245157	490314	817190	1144066	1352078	1352078
24	1	24	276	2024	10626	42504	134596	346104	735471	1307504	1961256	2496144	2704156
25	1	25	300	2300	12650	53130	177100	480700	1081575	2042975	3268760	4457400	5200300

For $r > 12$ use the relation:

$$\binom{n}{r} = \binom{n}{n-r}$$

Binomial coefficients arise in the expansion of $(1+x)^n$ in powers of x; they can be expressed in terms of factorials:

$$(1+x)^n = \sum_{r=0}^{n} \binom{n}{r} x^r$$

where

$$\binom{n}{r} = \frac{n!}{r!\,(n-r)!} = \frac{n(n-1)(n-2)\ldots(n-r+1)}{r!}$$

UPPER TAIL PROBABILITIES $Q(r)$ OF BINOMIAL DISTRIBUTIONS Bi(n, π)

									π							
n	r	.01	.02	.03	.04	.05	.06	.07	.08	.09	.10	.12	.14	.16	.18	.20
2	1	.0199	.0396	.0591	.0784	.0975	.1164	.1351	.1536	.1719	.1900	.2256	.2604	.2944	.3276	.3600
	2	.0001	.0004	.0009	.0016	.0025	.0036	.0049	.0064	.0081	.0100	.0144	.0196	.0256	.0324	.0400
3	1	.0297	.0588	.0873	.1153	.1426	.1694	.1956	.2213	.2464	.2710	.3185	.3639	.4073	.4486	.4880
	2	.0003	.0012	.0026	.0047	.0072	.0104	.0140	.0182	.0228	.0280	.0397	.0533	.0686	.0855	.1040
	3				.0001	.0001	.0002	.0003	.0005	.0007	.0010	.0017	.0027	.0041	.0058	.0080
4	1	.0394	.0776	.1147	.1507	.1855	.2193	.2519	.2836	.3143	.3439	.4003	.4530	.5021	.5479	.5904
	2	.0006	.0023	.0052	.0091	.0140	.0199	.0267	.0344	.0430	.0523	.0732	.0968	.1228	.1509	.1808
	3			.0001	.0002	.0005	.0008	.0013	.0019	.0027	.0037	.0063	.0098	.0144	.0202	.0272
	4									.0001	.0001	.0002	.0004	.0007	.0010	.0016
5	1	.0490	.0961	.1413	.1846	.2262	.2661	.3043	.3409	.3760	.4095	.4723	.5296	.5818	.6293	.6723
	2	.0010	.0038	.0085	.0148	.0226	.0319	.0425	.0544	.0674	.0815	.1125	.1467	.1835	.2224	.2627
	3		.0001	.0003	.0006	.0012	.0020	.0031	.0045	.0063	.0086	.0143	.0220	.0318	.0437	.0579
	4						.0001	.0001	.0002	.0003	.0005	.0009	.0017	.0029	.0045	.0067
	5												.0001	.0001	.0002	.0003
6	1	.0585	.1142	.1670	.2172	.2649	.3101	.3530	.3936	.4321	.4686	.5356	.5954	.6487	.6960	.7379
	2	.0015	.0057	.0125	.0216	.0328	.0459	.0608	.0773	.0952	.1143	.1556	.2003	.2472	.2956	.3446
	3		.0002	.0005	.0012	.0022	.0038	.0058	.0085	.0118	.0158	.0261	.0395	.0560	.0759	.0989
	4					.0001	.0002	.0003	.0005	.0008	.0013	.0025	.0045	.0075	.0116	.0170
	5										.0001	.0001	.0003	.0005	.0010	.0016
	6															.0001
7	1	.0679	.1319	.1920	.2486	.3017	.3515	.3983	.4422	.4832	.5217	.5913	.6521	.7049	.7507	.7903
	2	.0020	.0079	.0171	.0294	.0444	.0618	.0813	.1026	.1255	.1497	.2012	.2556	.3115	.3677	.4233
	3		.0003	.0009	.0020	.0038	.0063	.0097	.0140	.0193	.0257	.0416	.0620	.0866	.1154	.1480
	4				.0001	.0002	.0004	.0007	.0012	.0018	.0027	.0054	.0094	.0153	.0231	.0333
	5								.0001	.0001	.0002	.0004	.0009	.0017	.0029	.0047
	6													.0001	.0002	.0004
8	1	.0773	.1492	.2163	.2786	.3366	.3904	.4404	.4868	.5297	.5695	.6404	.7008	.7521	.7956	.8322
	2	.0027	.0103	.0223	.0381	.0572	.0792	.1035	.1298	.1577	.1869	.2480	.3111	.3744	.4366	.4967
	3	.0001	.0004	.0013	.0031	.0058	.0096	.0147	.0211	.0289	.0381	.0608	.0891	.1226	.1608	.2031
	4			.0001	.0002	.0004	.0007	.0013	.0022	.0034	.0050	.0097	.0168	.0267	.0397	.0563
	5							.0001	.0001	.0003	.0004	.0010	.0021	.0038	.0065	.0104
	6											.0001	.0002	.0004	.0007	.0012
	7															.0001
9	1	.0865	.1663	.2398	.3075	.3698	.4270	.4796	.5278	.5721	.6126	.6835	.7427	.7918	.8324	.8658
	2	.0034	.0131	.0282	.0478	.0712	.0978	.1271	.1583	.1912	.2252	.2951	.3657	.4348	.5012	.5638
	3	.0001	.0006	.0020	.0045	.0084	.0138	.0209	.0298	.0405	.0530	.0833	.1202	.1629	.2105	.2618
	4			.0001	.0003	.0006	.0013	.0023	.0037	.0057	.0083	.0158	.0269	.0420	.0615	.0856
	5					.0001	.0002	.0003	.0005	.0009	.0021	.0041	.0075	.0125	.0196
	6										.0001	.0002	.0004	.0009	.0017	.0031
	7													.0001	.0002	.0003
10	1	.0956	.1829	.2626	.3352	.4013	.4614	.5160	.5656	.6106	.6513	.7215	.7787	.8251	.8626	.8926
	2	.0043	.0162	.0345	.0582	.0861	.1176	.1517	.1879	.2254	.2639	.3417	.4184	.4920	.5608	.6242
	3	.0001	.0009	.0028	.0062	.0115	.0188	.0283	.0401	.0540	.0702	.1087	.1545	.2064	.2628	.3222
	4			.0001	.0004	.0010	.0020	.0036	.0058	.0088	.0128	.0239	.0400	.0614	.0883	.1209
	5					.0001	.0002	.0003	.0006	.0010	.0016	.0037	.0073	.0130	.0213	.0328
	6									.0001	.0001	.0004	.0010	.0020	.0037	.0064
	7												.0001	.0002	.0004	.0009
	8															.0001

For $r = 0$ the upper tail probabilities are all 1; they are omitted from the table. An entry $1-$ indicates a probability greater than 0.999 95. Where a space is left blank the probability is less than 0.000 05.

Source: For the tables on pp. 14–22, *Tables of the Cumulative Binomial Probability Distribution, Annals of the Harvard University Computation Laboratory*, vol. 35.

UPPER TAIL PROBABILITIES $Q(r)$ OF BINOMIAL DISTRIBUTIONS $Bi(n, \pi)$

n	r	.22	.24	.26	.28	.30	.32	.34	.36	.38	.40	.42	.44	.46	.48	.50
2	1	.3916	.4224	.4524	.4816	.5100	.5376	.5644	.5904	.6156	.6400	.6636	.6864	.7084	.7296	.7500
	2	.0484	.0576	.0676	.0784	.0900	.1024	.1156	.1296	.1444	.1600	.1764	.1936	.2116	.2304	.2500
3	1	.5254	.5610	.5948	.6268	.6570	.6856	.7125	.7379	.7617	.7840	.8049	.8244	.8425	.8594	.8750
	2	.1239	.1452	.1676	.1913	.2160	.2417	.2682	.2955	.3235	.3520	.3810	.4104	.4401	.4700	.5000
	3	.0106	.0138	.0176	.0220	.0270	.0328	.0393	.0467	.0549	.0640	.0741	.0852	.0973	.1106	.1250
4	1	.6298	.6664	.7001	.7313	.7599	.7862	.8103	.8322	.8522	.8704	.8868	.9017	.9150	.9269	.9375
	2	.2122	.2450	.2787	.3132	.3483	.3837	.4193	.4547	.4900	.5248	.5590	.5926	.6252	.6569	.6875
	3	.0356	.0453	.0566	.0694	.0837	.0996	.1171	.1362	.1569	.1792	.2030	.2283	.2550	.2831	.3125
	4	.0023	.0033	.0046	.0061	.0081	.0105	.0134	.0168	.0209	.0256	.0311	.0375	.0448	.0531	.0625
5	1	.7113	.7464	.7781	.8065	.8319	.8546	.8748	.8926	.9084	.9222	.9344	.9449	.9541	.9620	.9688
	2	.3041	.3461	.3883	.4303	.4718	.5125	.5522	.5906	.6276	.6630	.6967	.7286	.7585	.7865	.8125
	3	.0744	.0933	.1143	.1376	.1631	.1905	.2199	.2509	.2835	.3174	.3525	.3886	.4253	.4625	.5000
	4	.0097	.0134	.0181	.0238	.0308	.0390	.0486	.0598	.0726	.0870	.1033	.1214	.1415	.1635	.1875
	5	.0005	.0008	.0012	.0017	.0024	.0034	.0045	.0060	.0079	.0102	.0131	.0165	.0206	.0255	.0312
6	1	.7748	.8073	.8358	.8607	.8824	.9011	.9173	.9313	.9432	.9533	.9619	.9692	.9752	.9802	.9844
	2	.3937	.4422	.4896	.5356	.5798	.6220	.6619	.6994	.7343	.7667	.7965	.8238	.8485	.8707	.8906
	3	.1250	.1539	.1856	.2196	.2557	.2936	.3328	.3732	.4143	.4557	.4971	.5382	.5786	.6180	.6562
	4	.0239	.0326	.0431	.0557	.0705	.0875	.1069	.1286	.1527	.1792	.2080	.2390	.2721	.3070	.3438
	5	.0025	.0038	.0056	.0079	.0109	.0148	.0195	.0254	.0325	.0410	.0510	.0627	.0762	.0917	.1094
	6	.0001	.0002	.0003	.0005	.0007	.0011	.0015	.0022	.0030	.0041	.0055	.0073	.0095	.0122	.0156
7	1	.8243	.8535	.8785	.8997	.9176	.9328	.9454	.9560	.9648	.9720	.9779	.9827	.9866	.9897	.9922
	2	.4775	.5298	.5796	.6266	.6706	.7113	.7487	.7828	.8137	.8414	.8660	.8877	.9068	.9233	.9375
	3	.1841	.2231	.2646	.3081	.3529	.3987	.4447	.4906	.5359	.5801	.6229	.6638	.7027	.7393	.7734
	4	.0461	.0617	.0802	.1016	.1260	.1534	.1837	.2167	.2521	.2898	.3294	.3706	.4131	.4563	.5000
	5	.0072	.0107	.0153	.0213	.0288	.0380	.0492	.0625	.0782	.0963	.1169	.1402	.1663	.1951	.2266
	6	.0006	.0011	.0017	.0026	.0038	.0055	.0077	.0105	.0142	.0188	.0246	.0316	.0402	.0504	.0625
	7			.0001	.0001	.0002	.0003	.0005	.0008	.0011	.0016	.0023	.0032	.0044	.0059	.0078
8	1	.8630	.8887	.9101	.9278	.9424	.9543	.9640	.9719	.9782	.9832	.9872	.9903	.9928	.9947	.9961
	2	.5538	.6075	.6573	.7031	.7447	.7822	.8156	.8452	.8711	.8936	.9130	.9295	.9435	.9552	.9648
	3	.2486	.2967	.3465	.3973	.4482	.4987	.5481	.5958	.6415	.6846	.7250	.7624	.7966	.8276	.8555
	4	.0765	.1004	.1281	.1594	.1941	.2319	.2724	.3153	.3599	.4059	.4527	.4996	.5463	.5922	.6367
	5	.0158	.0230	.0322	.0438	.0580	.0750	.0949	.1180	.1443	.1737	.2062	.2416	.2798	.3205	.3633
	6	.0021	.0034	.0052	.0078	.0113	.0159	.0218	.0293	.0385	.0498	.0634	.0794	.0982	.1198	.1445
	7	.0002	.0003	.0005	.0008	.0013	.0020	.0030	.0043	.0061	.0085	.0117	.0157	.0208	.0272	.0352
	8					.0001	.0001	.0002	.0003	.0004	.0007	.0010	.0014	.0020	.0028	.0039
9	1	.8931	.9154	.9335	.9480	.9596	.9689	.9762	.9820	.9865	.9899	.9926	.9946	.9961	.9972	.9980
	2	.6218	.6750	.7230	.7660	.8040	.8372	.8661	.8908	.9118	.9295	.9442	.9563	.9662	.9741	.9805
	3	.3158	.3713	.4273	.4829	.5372	.5894	.6390	.6856	.7287	.7682	.8039	.8359	.8642	.8889	.9102
	4	.1144	.1475	.1849	.2260	.2703	.3173	.3662	.4163	.4669	.5174	.5670	.6152	.6614	.7052	.7461
	5	.0291	.0416	.0571	.0762	.0988	.1252	.1553	.1890	.2262	.2666	.3097	.3551	.4024	.4509	.5000
	6	.0051	.0081	.0122	.0179	.0253	.0348	.0467	.0612	.0787	.0994	.1233	.1508	.1817	.2161	.2539
	7	.0006	.0010	.0017	.0028	.0043	.0064	.0094	.0133	.0184	.0250	.0334	.0437	.0564	.0717	.0898
	8			.0001	.0001	.0003	.0004	.0007	.0011	.0017	.0026	.0038	.0055	.0077	.0107	.0145	.0195
	9								.0001	.0001	.0002	.0003	.0004	.0006	.0009	.0014	.0020
10	1	.9166	.9357	.9508	.9626	.9718	.9789	.9843	.9885	.9916	.9940	.9957	.9970	.9979	.9986	.9990
	2	.6815	.7327	.7778	.8170	.8507	.8794	.9035	.9236	.9402	.9536	.9645	.9731	.9799	.9852	.9893
	3	.3831	.4442	.5042	.5622	.6172	.6687	.7162	.7595	.7983	.8327	.8628	.8889	.9111	.9298	.9453
	4	.1587	.2012	.2479	.2979	.3504	.4044	.4589	.5132	.5664	.6177	.6665	.7123	.7547	.7933	.8281
	5	.0479	.0670	.0904	.1181	.1503	.1867	.2270	.2708	.3177	.3669	.4178	.4696	.5216	.5730	.6230
	6	.0104	.0161	.0239	.0342	.0473	.0637	.0836	.1072	.1348	.1662	.2016	.2407	.2832	.3288	.3770
	7	.0016	.0027	.0045	.0070	.0106	.0155	.0220	.0305	.0413	.0548	.0712	.0908	.1141	.1410	.1719
	8	.0002	.0003	.0006	.0010	.0016	.0025	.0039	.0059	.0086	.0123	.0172	.0236	.0317	.0420	.0547
	9				.0001	.0001	.0003	.0004	.0007	.0011	.0017	.0025	.0037	.0054	.0077	.0107
	10									.0001	.0001	.0002	.0003	.0004	.0006	.0010

UPPER TAIL PROBABILITIES $Q(r)$ OF BINOMIAL DISTRIBUTIONS Bi(n, π)

n	r	.01	.02	.03	.04	.05	.06	.07	.08 (π)	.09	.10	.12	.14	.16	.18	.20
11	1	.1047	.1993	.2847	.3618	.4312	.4937	.5499	.6004	.6456	.6862	.7549	.8097	.8531	.8873	.9141
	2	.0052	.0195	.0413	.0692	.1019	.1382	.1772	.2181	.2601	.3026	.3873	.4689	.5453	.6151	.6779
	3	.0002	.0012	.0037	.0083	.0152	.0248	.0370	.0519	.0695	.0896	.1366	.1915	.2521	.3164	.3826
	4			.0002	.0007	.0016	.0030	.0053	.0085	.0129	.0185	.0341	.0560	.0846	.1197	.1611
	5					.0001	.0003	.0005	.0010	.0017	.0028	.0061	.0119	.0207	.0334	.0504
	6								.0001	.0002	.0003	.0008	.0018	.0037	.0068	.0117
	7											.0001	.0002	.0005	.0010	.0020
	8														.0001	.0002
12	1	.1136	.2153	.3062	.3873	.4596	.5241	.5814	.6323	.6775	.7176	.7843	.8363	.8766	.9076	.9313
	2	.0062	.0231	.0486	.0809	.1184	.1595	.2033	.2487	.2948	.3410	.4314	.5166	.5945	.6641	.7251
	3	.0002	.0015	.0048	.0107	.0196	.0316	.0468	.0652	.0866	.1109	.1667	.2303	.2990	.3702	.4417
	4		.0001	.0003	.0010	.0022	.0043	.0075	.0120	.0180	.0256	.0464	.0750	.1114	.1552	.2054
	5				.0001	.0002	.0004	.0009	.0016	.0027	.0043	.0095	.0181	.0310	.0489	.0726
	6							.0001	.0002	.0003	.0005	.0014	.0033	.0065	.0116	.0194
	7										.0001	.0002	.0004	.0010	.0021	.0039
	8													.0001	.0003	.0006
	9															.0001
13	1	.1225	.2310	.3270	.4118	.4867	.5526	.6107	.6617	.7065	.7458	.8102	.8592	.8963	.9242	.9450
	2	.0072	.0270	.0564	.0932	.1354	.1814	.2298	.2794	.3293	.3787	.4738	.5614	.6396	.7080	.7664
	3	.0003	.0020	.0062	.0135	.0245	.0392	.0578	.0799	.1054	.1339	.1985	.2704	.3463	.4231	.4983
	4		.0001	.0005	.0014	.0031	.0060	.0103	.0163	.0242	.0342	.0609	.0967	.1414	.1939	.2527
	5				.0001	.0003	.0007	.0013	.0024	.0041	.0065	.0139	.0260	.0438	.0681	.0991
	6						.0001	.0001	.0003	.0005	.0009	.0024	.0053	.0104	.0183	.0300
	7										.0001	.0003	.0008	.0019	.0038	.0070
	8												.0001	.0003	.0006	.0012
	9														.0001	.0002
14	1	.1313	.2464	.3472	.4353	.5123	.5795	.6380	.6888	.7330	.7712	.8330	.8789	.9129	.9379	.9560
	2	.0084	.0310	.0645	.1059	.1530	.2037	.2564	.3100	.3632	.4154	.5141	.6031	.6807	.7469	.8021
	3	.0003	.0025	.0077	.0167	.0301	.0478	.0698	.0958	.1255	.1584	.2315	.3111	.3932	.4744	.5519
	4		.0001	.0006	.0019	.0042	.0080	.0136	.0214	.0315	.0441	.0774	.1210	.1742	.2351	.3018
	5				.0002	.0004	.0010	.0020	.0035	.0059	.0092	.0196	.0359	.0594	.0907	.1298
	6						.0001	.0002	.0004	.0008	.0015	.0038	.0082	.0157	.0273	.0439
	7									.0001	.0002	.0006	.0015	.0032	.0064	.0116
	8											.0001	.0002	.0005	.0012	.0024
	9													.0001	.0002	.0004
15	1	.1399	.2614	.3667	.4579	.5367	.6047	.6633	.7137	.7570	.7941	.8530	.8959	.9269	.9490	.9648
	2	.0096	.0353	.0730	.1191	.1710	.2262	.2832	.3403	.3965	.4510	.5524	.6417	.7179	.7813	.8329
	3	.0004	.0030	.0094	.0203	.0362	.0571	.0829	.1130	.1469	.1841	.2654	.3520	.4392	.5234	.6020
	4		.0002	.0008	.0024	.0055	.0104	.0175	.0273	.0399	.0556	.0959	.1476	.2092	.2782	.3518
	5			.0001	.0002	.0006	.0014	.0028	.0050	.0082	.0127	.0265	.0478	.0778	.1167	.1642

If the probability of success in a single trial is π, the probability $p(r)$ of exactly r successes in n independent trials is given by the binomial distribution Bi(n, π) (or B(n, π)):

$$p(r) = \binom{n}{r}\pi^r(1-\pi)^{n-r} \quad (r = 0, 1, 2, \ldots, n)$$

The mean of the distribution is $n\pi$ and the variance is $n\pi(1-\pi)$.

The table gives, for $r > 0$, the probability $Q(r)$ of r or more successes: $Q(r) = \sum_{x=r}^{n} p(x)$. The individual terms can be found by subtraction:

$$p(r) = Q(r) - Q(r+1) \quad (r < n), \quad p(n) = Q(n)$$

The probability $P(r)$ of r or fewer successes (*the cumulative probability*) is given by:

$$P(r) = 1 - Q(r+1) \quad (r < n), \quad P(n) = 1$$

UPPER TAIL PROBABILITIES $Q(r)$ OF BINOMIAL DISTRIBUTIONS Bi(n, π)

n	r	.22	.24	.26	.28	.30	.32	.34	.36 (π)	.38	.40	.42	.44	.46	.48	.50
11	1	.9350	.9511	.9636	.9730	.9802	.9856	.9896	.9926	.9948	.9964	.9975	.9983	.9989	.9992	.9995
	2	.7333	.7814	.8227	.8577	.8870	.9112	.9310	.9470	.9597	.9698	.9776	.9836	.9882	.9916	.9941
	3	.4488	.5134	.5753	.6335	.6873	.7361	.7799	.8186	.8522	.8811	.9055	.9260	.9428	.9564	.9673
	4	.2081	.2596	.3146	.3719	.4304	.4890	.5464	.6019	.6545	.7037	.7490	.7900	.8266	.8588	.8867
	5	.0723	.0992	.1313	.1685	.2103	.2563	.3059	.3581	.4122	.4672	.5223	.5764	.6288	.6787	.7256
	6	.0186	.0283	.0412	.0577	.0782	.1031	.1324	.1661	.2043	.2465	.2924	.3414	.3929	.4460	.5000
	7	.0035	.0059	.0095	.0146	.0216	.0309	.0430	.0581	.0768	.0994	.1260	.1568	.1919	.2312	.2744
	8	.0005	.0009	.0016	.0027	.0043	.0067	.0101	.0148	.0210	.0293	.0399	.0532	.0696	.0895	.1133
	9		.0001	.0002	.0003	.0006	.0010	.0016	.0026	.0039	.0059	.0087	.0125	.0175	.0241	.0327
	10						.0001	.0002	.0003	.0005	.0007	.0012	.0018	.0027	.0040	.0059
	11											.0001	.0001	.0002	.0003	.0005
12	1	.9493	.9629	.9730	.9806	.9862	.9902	.9932	.9953	.9968	.9978	.9986	.9990	.9994	.9996	.9998
	2	.7776	.8222	.8594	.8900	.9150	.9350	.9509	.9634	.9730	.9804	.9860	.9901	.9931	.9953	.9968
	3	.5114	.5778	.6397	.6963	.7472	.7922	.8313	.8648	.8931	.9166	.9358	.9513	.9637	.9733	.9807
	4	.2610	.3205	.3824	.4452	.5075	.5681	.6258	.6799	.7296	.7747	.8147	.8498	.8801	.9057	.9270
	5	.1021	.1377	.1790	.2254	.2763	.3308	.3876	.4459	.5043	.5618	.6175	.6704	.7198	.7652	.8062
	6	.0304	.0453	.0646	.0887	.1178	.1521	.1913	.2352	.2833	.3348	.3889	.4448	.5014	.5577	.6128
	7	.0068	.0113	.0178	.0267	.0386	.0540	.0734	.0970	.1253	.1582	.1959	.2380	.2843	.3343	.3872
	8	.0011	.0021	.0036	.0060	.0095	.0144	.0213	.0304	.0422	.0573	.0760	.0988	.1258	.1575	.1938
	9	.0001	.0003	.0005	.0010	.0017	.0028	.0045	.0070	.0104	.0153	.0218	.0304	.0415	.0555	.0730
	10			.0001	.0001	.0002	.0004	.0007	.0011	.0018	.0028	.0043	.0065	.0095	.0137	.0193
	11							.0001	.0001	.0002	.0003	.0005	.0009	.0014	.0021	.0032
	12											.0001	.0001	.0001	.0001	.0002
13	1	.9604	.9718	.9800	.9860	.9903	.9934	.9955	.9970	.9980	.9987	.9992	.9995	.9997	.9998	.9999
	2	.8154	.8559	.8889	.9154	.9363	.9527	.9653	.0749	.9821	.9874	.9912	.9940	.9960	.9974	.9983
	3	.5699	.6364	.6968	.7505	.7975	.8379	.8720	.9003	.9235	.9421	.9569	.9684	.9772	.9838	.9888
	4	.3161	.3822	.4493	.5155	.5794	.6398	.6957	.7464	.7917	.8314	.8656	.8945	.9185	.9381	.9539
	5	.1371	.1816	.2319	.2870	.3457	.4067	.4686	.5301	.5899	.6470	.7003	.7493	.7935	.8326	.8666
	6	.0462	.0675	.0944	.1270	.1654	.2093	.2581	.3111	.3673	.4256	.4849	.5441	.6019	.6573	.7095
	7	.0120	.0195	.0299	.0440	.0624	.0854	.1135	.1468	.1853	.2288	.2770	.3290	.3842	.4415	.5000
	8	.0024	.0043	.0073	.0118	.0182	.0271	.0390	.0544	.0738	.0977	.1264	.1600	.1988	.2424	.2905
	9	.0004	.0007	.0013	.0024	.0040	.0065	.0102	.0154	.0225	.0321	.0446	.0605	.0803	.1045	.1334
	10		.0001	.0002	.0004	.0007	.0012	.0020	.0032	.0051	.0078	.0117	.0170	.0242	.0338	.0461
	11					.0001	.0001	.0003	.0005	.0008	.0013	.0021	.0033	.0051	.0077	.0112
	12									.0001	.0001	.0002	.0004	.0007	.0011	.0017
	13													.0001	.0001	.0001
14	1	.9691	.9786	.9852	.9899	.9932	.9955	.9970	.9981	.9988	.9992	.9995	.9997	.9998	.9999	.9999
	2	.8473	.8837	.9126	.9352	.9525	.9657	.9756	.9828	.9881	.9919	.9946	.9964	.9977	.9985	.9991
	3	.6239	.6891	.7467	.7967	.8392	.8746	.9037	.9271	.9457	.9602	.9713	.9797	.9858	.9903	.9935
	4	.3719	.4432	.5136	.5813	.6448	.7032	.7556	.8018	.8418	.8757	.9039	.9270	.9455	.9601	.9713
	5	.1765	.2297	.2884	.3509	.4158	.4813	.5458	.6080	.6666	.7207	.7697	.8132	.8510	.8833	.9102
	6	.0662	.0949	.1301	.1718	.2195	.2724	.3297	.3899	.4519	.5141	.5754	.6344	.6900	.7415	.7880
	7	.0196	.0310	.0467	.0673	.0933	.1250	.1626	.2059	.2545	.3075	.3643	.4236	.4843	.5451	.6047
	8	.0045	.0079	.0132	.0208	.0315	.0458	.0643	.0876	.1162	.1501	.1896	.2344	.2840	.3380	.3953
	9	.0008	.0016	.0029	.0050	.0083	.0131	.0200	.0294	.0420	.0583	.0789	.1043	.1348	.1707	.2120
	10	.0001	.0002	.0005	.0009	.0017	.0029	.0048	.0076	.0117	.0175	.0255	.0361	.0500	.0677	.0898
	11			.0001	.0001	.0002	.0005	.0008	.0014	.0024	.0039	.0061	.0093	.0139	.0202	.0287
	12						.0001	.0001	.0002	.0003	.0006	.0010	.0017	.0027	.0042	.0065
	13										.0001	.0001	.0002	.0003	.0006	.0009
	14															.0001
15	1	.9759	.9837	.9891	.9928	.9953	.9969	.9980	.9988	.9992	.9995	.9997	.9998	.9999	.9999	1 −
	2	.8741	.9065	.9315	.9505	.9647	.9752	.9829	.9883	.9922	.9948	.9966	.9979	.9987	.9992	.9995
	3	.6731	.7358	.7899	.8355	.8732	.9038	.9281	.9472	.9618	.9729	.9811	.9870	.9913	.9943	.9963
	4	.4274	.5022	.5742	.6416	.7031	.7580	.8060	.8469	.8813	.9095	.9322	.9502	.9641	.9746	.9824
	5	.2195	.2810	.3469	.4154	.4845	.5523	.6171	.6778	.7332	.7827	.8261	.8633	.8945	.9201	.9408

Values of π are given in the column headings.

n	r	.01	.02	.03	.04	.05	.06	.07	.08	.09	.10	.12	.14	.16	.18	.20
15	6					.0001	.0001	.0003	.0007	.0013	.0022	.0057	.0121	.0227	.0387	.0611
	7								.0001	.0002	.0003	.0010	.0024	.0052	.0102	.0181
	8											.0001	.0004	.0010	.0021	.0042
	9													.0001	.0003	.0008
	10															.0001
16	1	.1485	.2762	.3857	.4796	.5599	.6284	.6869	.7366	.7789	.8147	.8707	.9105	.9386	.9582	.9719
	2	.0109	.0399	.0818	.1327	.1892	.2489	.3098	.3701	.4289	.4853	.5885	.6773	.7513	.8115	.8593
	3	.0005	.0037	.0113	.0242	.0429	.0673	.0969	.1311	.1694	.2108	.2999	.3926	.4838	.5698	.6482
	4		.0002	.0011	.0032	.0070	.0132	.0221	.0342	.0496	.0684	.1162	.1763	.2460	.3223	.4019
	5			.0001	.0003	.0009	.0019	.0038	.0068	.0111	.0170	.0348	.0618	.0988	.1458	.2018
	6					.0001	.0002	.0005	.0010	.0019	.0033	.0082	.0171	.0315	.0527	.0817
	7							.0001	.0001	.0003	.0005	.0015	.0038	.0080	.0153	.0267
	8										.0001	.0002	.0007	.0016	.0036	.0070
	9												.0001	.0003	.0007	.0015
	10													.0001	.0001	.0002
17	1	.1571	.2907	.4042	.5004	.5819	.6507	.7088	.7577	.7988	.8332	.8862	.9230	.9484	.9657	.9775
	2	.0123	.0446	.0909	.1465	.2078	.2717	.3362	.3995	.4604	.5182	.6223	.7099	.7813	.8379	.8818
	3	.0006	.0044	.0134	.0286	.0503	.0782	.1118	.1503	.1927	.2382	.3345	.4324	.5266	.6133	.6904
	4		.0003	.0014	.0040	.0088	.0164	.0273	.0419	.0603	.0826	.1383	.2065	.2841	.3669	.4511
	5			.0001	.0004	.0012	.0026	.0051	.0090	.0145	.0221	.0446	.0778	.1224	.1775	.2418
	6					.0001	.0003	.0007	.0015	.0027	.0047	.0114	.0234	.0423	.0695	.1057
	7							.0001	.0002	.0004	.0008	.0023	.0056	.0118	.0220	.0377
	8										.0001	.0004	.0011	.0027	.0057	.0109
	9											.0001	.0002	.0005	.0012	.0026
	10												.0001	.0002	.0002	.0005
	11															.0001
18	1	.1655	.3049	.4220	.5204	.6028	.6717	.7292	.7771	.8169	.8499	.8998	.9338	.9566	.9719	.9820
	2	.0138	.0495	.1003	.1607	.2265	.2945	.3622	.4281	.4909	.5497	.6540	.7398	.8080	.8609	.9009
	3	.0007	.0052	.0157	.0333	.0581	.0898	.1275	.1702	.2168	.2662	.3690	.4713	.5673	.6538	.7287
	4		.0004	.0018	.0050	.0109	.0201	.0333	.0506	.0723	.0982	.1618	.2382	.3229	.4112	.4990
	5			.0002	.0006	.0015	.0034	.0067	.0116	.0187	.0282	.0558	.0959	.1482	.2116	.2836
	6				.0001	.0002	.0005	.0010	.0021	.0038	.0064	.0154	.0310	.0551	.0889	.1329
	7							.0001	.0003	.0006	.0012	.0034	.0081	.0167	.0306	.0513
	8									.0001	.0002	.0006	.0017	.0041	.0086	.0163
	9											.0001	.0003	.0008	.0020	.0043
	10													.0001	.0004	.0009
	11														.0001	.0002

Linear interpolation with respect to π should be carried out cautiously, since errors may in places amount to several units in the third decimal place.

π greater than $\frac{1}{2}$. Since the probability of r or more successes is equal to the probability of $n-r$ or fewer failures

$$Q(r|n, \pi) = P(n-r|n,\ 1-\pi) = 1 - Q(n-r+1|n,\ 1-\pi)$$

For $\pi > \frac{1}{2}$, $Q(n-r+1|n,\ 1-\pi)$ can be read from the tables. For example, the probability of 4 or more successes in 6 independent trials with $\pi = 0.7$ is $1 - Q(6-4+1|6, 0.3)$; its value is $1 - 0.2557$ or 0.7443.

Sampling distribution of the number of successes.

A binomial distribution can be regarded as the sampling distribution of the number r of successes in a random sample of size n drawn from a large population in which the proportion of successes is π. The sample proportion $p = r/n$ is an unbiased estimator of π (i.e. the expected value of p is π).

UPPER TAIL PROBABILITIES $Q(r)$ OF BINOMIAL DISTRIBUTIONS Bi(n, π)

n r	.22	.24	.26	.28	.30	.32	.34	π .36	.38	.40	.42	.44	.46	.48	.50
15 6	.0905	.1272	.1713	.2220	.2784	.3393	.4032	.4684	.5335	.5968	.6570	.7131	.7641	.8095	.8491
7	.0298	.0463	.0684	.0965	.1311	.1722	.2194	.2722	.3295	.3902	.4530	.5164	.5789	.6394	.6964
8	.0078	.0135	.0219	.0338	.0500	.0711	.0977	.1302	.1687	.2131	.2630	.3176	.3762	.4374	.5000
9	.0016	.0031	.0056	.0094	.0152	.0236	.0351	.0504	.0702	.0950	.1254	.1615	.2034	.2510	.3036
10	.0003	.0006	.0011	.0021	.0037	.0062	.0099	.0154	.0232	.0338	.0479	.0661	.0890	.1171	.1509
11		.0001	.0002	.0003	.0007	.0012	.0022	.0037	.0059	.0093	.0143	.0211	.0305	.0430	.0592
12					.0001	.0002	.0004	.0006	.0011	.0019	.0032	.0051	.0079	.0119	.0176
13								.0001	.0002	.0003	.0005	.0009	.0014	.0023	.0037
14												.0001	.0002	.0003	.0005
16 1	.9812	.9876	.9919	.9948	.9967	.9979	.9987	.9992	.9995	.9997	.9998	.9999	.9999	1 −	1 −
2	.8965	.9250	.9465	.9623	.9739	.9822	.9880	.9921	.9948	.9967	.9979	.9987	.9992	.9995	.9997
3	.7173	.7768	.8267	.8677	.9006	.9266	.9467	.9620	.9734	.9817	.9876	.9918	.9947	.9966	.9979
4	.4814	.5583	.6303	.6959	.7541	.8047	.8475	.8830	.9119	.9349	.9527	.9664	.9766	.9840	.9894
5	.2652	.3341	.4060	.4788	.5501	.6181	.6813	.7387	.7895	.8334	.8707	.9015	.9265	.9463	.9616
6	.1188	.1641	.2169	.2761	.3402	.4074	.4759	.5438	.6094	.6712	.7280	.7792	.8241	.8626	.8949
7	.0432	.0658	.0951	.1317	.1753	.2257	.2819	.3428	.4070	.4728	.5387	.6029	.6641	.7210	.7728
8	.0127	.0214	.0340	.0514	.0744	.1035	.1391	.1813	.2298	.2839	.3428	.4051	.4694	.5343	.5982
9	.0030	.0056	.0098	.0163	.0257	.0388	.0564	.0791	.1076	.1423	.1832	.2302	.2829	.3405	.4018
10	.0006	.0012	.0023	.0041	.0071	.0117	.0185	.0280	.0411	.0583	.0805	.1081	.1416	.1814	.2272
11	.0001	.0002	.0004	.0008	.0016	.0028	.0048	.0079	.0125	.0191	.0284	.0409	.0574	.0786	.1051
12			.0001	.0001	.0003	.0005	.0010	.0017	.0030	.0049	.0078	.0121	.0183	.0268	.0384
13						.0001	.0001	.0003	.0005	.0009	.0016	.0027	.0044	.0069	.0106
14									.0001	.0001	.0002	.0004	.0007	.0013	.0021
15													.0001	.0001	.0003
17 1	.9854	.9906	.9940	.9962	.9977	.9986	.9991	.9995	.9997	.9998	.9999	.9999	1 −	1 −	1 −
2	.9152	.9400	.9583	.9714	.9807	.9872	.9917	.9946	.9966	.9979	.9987	.9992	.9996	.9998	.9999
3	.7567	.8123	.8578	.8942	.9226	.9444	.9608	.9728	.9815	.9877	.9920	.9948	.9968	.9980	.9988
4	.5333	.6107	.6814	.7440	.7981	.8437	.8812	.9115	.9353	.9536	.9674	.9776	.9849	.9901	.9936
5	.3128	.3879	.4643	.5396	.6113	.6778	.7378	.7906	.8360	.8740	.9051	.9301	.9495	.9644	.9755
6	.1510	.2049	.2661	.3329	.4032	.4749	.5458	.6139	.6778	.7361	.7879	.8330	.8712	.9028	.9283
7	.0598	.0894	.1268	.1721	.2248	.2838	.3479	.4152	.4839	.5522	.6182	.6805	.7377	.7890	.8338
8	.0194	.0320	.0499	.0739	.1046	.1426	.1877	.2395	.2971	.3595	.4250	.4921	.5590	.6239	.6855
9	.0051	.0094	.0161	.0261	.0403	.0595	.0845	.1159	.1541	.1989	.2502	.3072	.3687	.4335	.5000
10	.0011	.0022	.0042	.0075	.0127	.0204	.0314	.0464	.0664	.0919	.1236	.1618	.2066	.2577	.3145
11	.0002	.0004	.0009	.0018	.0032	.0057	.0095	.0151	.0234	.0348	.0503	.0705	.0962	.1279	.1662
12		.0001	.0002	.0003	.0007	.0013	.0023	.0040	.0066	.0106	.0165	.0248	.0363	.0517	.0717
13					.0001	.0002	.0004	.0008	.0015	.0025	.0042	.0069	.0108	.0165	.0245
14							.0001	.0001	.0002	.0005	.0008	.0014	.0024	.0040	.0064
15										.0001	.0001	.0002	.0004	.0007	.0012
16														.0001	.0001
18 1	.9886	.9928	.9956	.9973	.9984	.9990	.9994	.9997	.9998	.9999	.9999	1 −	1 −	1 −	1 −
2	.9306	.9522	.9676	.9784	.9858	.9908	.9942	.9964	.9978	.9987	.9992	.9996	.9998	.9999	.9999
3	.7916	.8430	.8839	.9158	.9400	.9581	.9713	.9807	.9873	.9918	.9948	.9968	.9981	.9989	.9993
4	.5825	.6591	.7272	.7860	.8354	.8759	.9083	.9335	.9528	.9672	.9777	.9852	.9904	.9939	.9962
5	.3613	.4414	.5208	.5968	.6673	.7309	.7866	.8341	.8737	.9058	.9313	.9510	.9658	.9767	.9846
6	.1866	.2488	.3176	.3907	.4656	.5398	.6111	.6776	.7379	.7912	.8372	.8757	.9072	.9324	.9519
7	.0799	.1171	.1630	.2171	.2783	.3450	.4151	.4867	.5576	.6257	.6895	.7476	.7991	.8436	.8811
8	.0283	.0458	.0699	.1014	.1407	.1878	.2421	.3027	.3681	.4366	.5062	.5750	.6412	.7032	.7597
9	.0083	.0148	.0249	.0395	.0596	.0861	.1196	.1604	.2084	.2632	.3236	.3885	.4562	.5249	.5927
10	.0020	.0039	.0073	.0127	.0210	.0329	.0494	.0714	.0997	.1347	.1768	.2258	.2812	.3421	.4073
11	.0004	.0009	.0018	.0034	.0061	.0104	.0169	.0264	.0397	.0576	.0811	.1107	.1470	.1902	.2403
12	.0001	.0002	.0003	.0007	.0014	.0027	.0047	.0080	.0130	.0203	.0307	.0449	.0638	.0883	.1189
13			.0001	.0001	.0003	.0005	.0011	.0019	.0034	.0058	.0094	.0147	.0225	.0334	.0481
14						.0001	.0002	.0004	.0007	.0013	.0022	.0038	.0063	.0100	.0154
15								.0001	.0001	.0002	.0004	.0007	.0013	.0023	.0038
16												.0001	.0001	.0002	.0004	.0007
17															.0001

UPPER TAIL PROBABILITIES $Q(r)$ OF BINOMIAL DISTRIBUTIONS Bi(n, π)

n	r	.01	.02	.03	.04	.05	.06	.07	.08	.09	.10	.12	.14	.16	.18	.20
19	1	.1738	.3188	.4394	.5396	.6226	.6914	.7481	.7949	.8334	.8649	.9119	.9431	.9636	.9770	.9856
	2	.0153	.0546	.1100	.1751	.2453	.3171	.3879	.4560	.5202	.5797	.6835	.7669	.8318	.8809	.9171
	3	.0009	.0061	.0183	.0384	.0665	.1021	.1439	.1908	.2415	.2946	.4032	.5089	.6059	.6910	.7631
	4		.0005	.0022	.0061	.0132	.0243	.0398	.0602	.0853	.1150	.1867	.2708	.3620	.4549	.5449
	5			.0002	.0007	.0020	.0044	.0085	.0147	.0235	.0352	.0685	.1158	.1762	.2476	.3267
	6				.0001	.0002	.0006	.0014	.0029	.0051	.0086	.0202	.0401	.0700	.1110	.1631
	7						.0001	.0002	.0004	.0009	.0017	.0048	.0113	.0228	.0411	.0676
	8								.0001	.0001	.0003	.0009	.0026	.0061	.0126	.0233
	9											.0002	.0005	.0014	.0032	.0067
	10												.0001	.0002	.0007	.0016
	11														.0001	.0003
	12															.0001
20	1	.1821	.3324	.4562	.5580	.6415	.7099	.7658	.8113	.8484	.8784	.9224	.9510	.9694	.9811	.9885
	2	.0169	.0599	.1198	.1897	.2642	.3395	.4131	.4831	.5484	.6083	.7109	.7916	.8529	.8982	.9308
	3	.0010	.0071	.0210	.0439	.0755	.1150	.1610	.2121	.2666	.3231	.4369	.5450	.6420	.7252	.7939
	4		.0006	.0027	.0074	.0159	.0290	.0471	.0706	.0993	.1330	.2127	.3041	.4010	.4974	.5886
	5			.0003	.0010	.0026	.0056	.0107	.0183	.0290	.0432	.0827	.1375	.2059	.2849	.3704
	6				.0001	.0003	.0009	.0019	.0038	.0068	.0113	.0260	.0507	.0870	.1356	.1958
	7						.0001	.0003	.0006	.0013	.0024	.0067	.0153	.0304	.0537	.0867
	8								.0001	.0002	.0004	.0014	.0038	.0088	.0177	.0321
	9										.0001	.0002	.0008	.0021	.0049	.0100
	10												.0001	.0004	.0011	.0026
	11													.0001	.0002	.0006
	12															.0001
25	1	.2222	.3965	.5330	.6396	.7226	.7871	.8370	.8756	.9054	.9282	.9591	.9770	.9872	.9930	.9962
	2	.0258	.0886	.1720	.2642	.3576	.4473	.5304	.6053	.6714	.7288	.8195	.8832	.9263	.9546	.9726
	3	.0020	.0132	.0380	.0765	.1271	.1871	.2534	.3232	.3937	.4629	.5912	.7000	.7870	.8533	.9018
	4	.0001	.0014	.0062	.0165	.0341	.0598	.0936	.1351	.1831	.2364	.3525	.4714	.5837	.6829	.7660
	5		.0001	.0008	.0028	.0072	.0150	.0274	.0451	.0686	.0980	.1734	.2668	.3707	.4772	.5793
	6			.0001	.0004	.0012	.0031	.0065	.0123	.0210	.0334	.0709	.1268	.2002	.2875	.3833
	7					.0002	.0005	.0013	.0028	.0054	.0095	.0243	.0509	.0920	.1488	.2200
	8						.0001	.0002	.0005	.0011	.0023	.0070	.0173	.0361	.0661	.1091
	9								.0001	.0002	.0005	.0017	.0050	.0121	.0252	.0468
	10										.0001	.0004	.0013	.0035	.0083	.0173
	11											.0001	.0003	.0009	.0024	.0056
	12													.0002	.0006	.0015
	13														.0001	.0004
	14															.0001

The negative binomial distribution. If the probability of success in a single trial is π, the probability $q(m)$ that the kth success occurs at the mth trial in a sequence of independent trials is given, for fixed k, by the negative binomial distribution:

$$q(m) = \binom{m-1}{k-1} \pi^k (1-\pi)^{m-k} \quad (m = k, k+1, k+2, \ldots)$$

The mean of m is k/π and the variance is $k(1-\pi)/\pi^2$. The cumulative probability that the kth success occurs at the mth trial or earlier is equal to the upper tail probability $Q(k|m, \pi)$ for the binomial distribution Bi(m, π) and may be read from the tables.

UPPER TAIL PROBABILITIES $Q(r)$ OF BINOMIAL DISTRIBUTIONS Bi(n, π)

n	r	.22	.24	.26	.28	.30	.32	.34	.36	.38	.40	.42	.44	.46	.48	.50
19	1	.9911	.9946	.9967	.9981	.9989	.9993	.9996	.9998	.9999	.9999	1−	1−	1−	1−	
	2	.9434	.9619	.9749	.9837	.9896	.9935	.9960	.9976	.9986	.9992	.9995	.9997	.9999	.9999	1−
	3	.8222	.8692	.9057	.9333	.9538	.9686	.9791	.9863	.9913	.9945	.9967	.9980	.9988	.9993	.9996
	4	.6285	.7032	.7680	.8224	.8668	.9022	.9297	.9505	.9659	.9770	.9849	.9903	.9939	.9963	.9978
	5	.4100	.4936	.5744	.6498	.7178	.7773	.8280	.8699	.9038	.9304	.9508	.9660	.9771	.9850	.9904
	6	.2251	.2950	.3705	.4484	.5261	.6010	.6707	.7339	.7895	.8371	.8767	.9088	.9342	.9537	.9682
	7	.1034	.1487	.2032	.2657	.3345	.4073	.4818	.5554	.6261	.6919	.7515	.8039	.8488	.8862	.9165
	8	.0396	.0629	.0941	.1338	.1820	.2381	.3010	.3690	.4401	.5122	.5832	.6509	.7138	.7706	.8204
	9	.0127	.0222	.0366	.0568	.0839	.1186	.1612	.2116	.2691	.3325	.4003	.4706	.5413	.6105	.6762
	10	.0034	.0066	.0119	.0202	.0326	.0499	.0733	.1035	.1410	.1861	.2385	.2974	.3617	.4299	.5000
	11	.0007	.0016	.0032	.0060	.0105	.0176	.0280	.0426	.0625	.0885	.1213	.1613	.2087	.2631	.3238
	12	.0001	.0003	.0007	.0015	.0028	.0051	.0089	.0146	.0231	.0352	.0518	.0738	.1021	.1372	.1796
	13		.0001	.0001	.0003	.0006	.0012	.0023	.0041	.0070	.0116	.0183	.0280	.0415	.0597	.0835
	14					.0001	.0002	.0005	.0009	.0017	.0031	.0052	.0086	.0137	.0212	.0318
	15							.0001	.0002	.0003	.0006	.0012	.0021	.0036	.0060	.0096
	16										.0001	.0002	.0004	.0007	.0013	.0022
	17												.0001	.0001	.0002	.0004
20	1	.9931	.9959	.9976	.9986	.9992	.9996	.9998	.9999	.9999	1−	1−	1−	1−		
	2	.9539	.9698	.9805	.9877	.9924	.9953	.9972	.9984	.9991	.9995	.9997	.9998	.9999	1−	1−
	3	.8488	.8915	.9237	.9474	.9645	.9765	.9848	.9904	.9940	.9964	.9979	.9988	.9993	.9996	.9998
	4	.6711	.7431	.8038	.8534	.8929	.9235	.9465	.9634	.9755	.9840	.9898	.9937	.9962	.9977	.9987
	5	.4580	.5439	.6248	.6981	.7625	.8173	.8626	.8989	.9274	.9490	.9651	.9767	.9848	.9904	.9941
	6	.2657	.3427	.4235	.5048	.5836	.6574	.7242	.7829	.8329	.8744	.9078	.9340	.9539	.9687	.9793
	7	.1301	.1838	.2467	.3169	.3920	.4693	.5460	.6197	.6882	.7500	.8041	.8501	.8881	.9186	.9423
	8	.0536	.0835	.1225	.1707	.2277	.2922	.3624	.4361	.5108	.5841	.6539	.7183	.7759	.8261	.8684
	9	.0186	.0320	.0515	.0784	.1133	.1568	.2087	.2683	.3341	.4044	.4771	.5499	.6207	.6873	.7483
	10	.0054	.0103	.0183	.0305	.0480	.0719	.1032	.1424	.1897	.2447	.3064	.3736	.4443	.5166	.5881
	11	.0013	.0028	.0055	.0100	.0171	.0279	.0434	.0645	.0923	.1275	.1705	.2212	.2791	.3432	.4119
	12	.0003	.0006	.0014	.0027	.0051	.0091	.0154	.0247	.0381	.0565	.0810	.1123	.1511	.1977	.2517
	13		.0001	.0003	.0006	.0013	.0025	.0045	.0079	.0132	.0210	.0324	.0482	.0694	.0969	.1316
	14				.0001	.0003	.0006	.0011	.0021	.0037	.0065	.0107	.0172	.0265	.0397	.0577
	15						.0001	.0002	.0004	.0009	.0016	.0029	.0050	.0083	.0133	.0207
	16								.0001	.0002	.0003	.0006	.0011	.0020	.0035	.0059
	17										.0001	.0002	.0004	.0007	.0013
	18												.0001	.0001	.0002
25	1	.9980	.9990	.9995	.9997	.9999	.9999	1−	1−	1−	1−					
	2	.9838	.9907	.9947	.9971	.9984	.9992	.9996	.9998	.9999	.9999	1−	1−			
	3	.9360	.9593	.9748	.9848	.9910	.9949	.9971	.9984	.9992	.9996	.9998	.9999	1−	1−	1−
	4	.8324	.8834	.9211	.9481	.9668	.9793	.9874	.9926	.9958	.9976	.9987	.9993	.9997	.9998	.9999
	5	.6718	.7516	.8174	.8696	.9095	.9390	.9600	.9745	.9842	.9905	.9945	.9969	.9983	.9991	.9995
	6	.4816	.5767	.6644	.7415	.8065	.8593	.9006	.9318	.9546	.9706	.9816	.9888	.9934	.9963	.9980
	7	.3027	.3927	.4851	.5753	.6593	.7343	.7987	.8517	.8940	.9264	.9505	.9677	.9796	.9876	.9927
	8	.1658	.2349	.3142	.3999	.4882	.5747	.6561	.7295	.7932	.8464	.8894	.9227	.9477	.9658	.9784
	9	.0788	.1228	.1790	.2465	.3231	.4057	.4908	.5748	.6542	.7265	.7897	.8431	.8865	.9205	.9461
	10	.0325	.0560	.0893	.1338	.1894	.2555	.3300	.4104	.4933	.5754	.6535	.7250	.7880	.8415	.8852
	11	.0117	.0222	.0389	.0636	.0978	.1424	.1975	.2624	.3355	.4142	.4956	.5765	.6538	.7249	.7878
	12	.0036	.0076	.0148	.0264	.0442	.0698	.1044	.1490	.2036	.2677	.3397	.4174	.4978	.5780	.6550
	13	.0010	.0023	.0049	.0096	.0175	.0299	.0485	.0745	.1093	.1538	.2080	.2715	.3429	.4199	.5000
	14	.0002	.0006	.0014	.0030	.0060	.0112	.0196	.0326	.0515	.0778	.1127	.1569	.2109	.2740	.3450
	15		.0001	.0003	.0008	.0018	.0036	.0069	.0124	.0212	.0344	.0535	.0797	.1145	.1585	.2122
	16			.0001	.0002	.0005	.0010	.0021	.0041	.0075	.0132	.0220	.0353	.0543	.0803	.1148
	17					.0001	.0002	.0005	.0011	.0023	.0043	.0078	.0134	.0222	.0352	.0539
	18							.0001	.0003	.0006	.0012	.0023	.0044	.0077	.0132	.0216
	19								.0001	.0001	.0003	.0006	.0012	.0023	.0041	.0073
	20										.0001	.0001	.0003	.0005	.0011	.0020
	21													.0001	.0002	.0005
	22															.0001

UPPER TAIL PROBABILITIES $Q(r)$ OF BINOMIAL DISTRIBUTIONS Bi(n, π)

n	r	$\pi=\frac{1}{8}$	$\frac{1}{6}$	$\frac{1}{4}$	$\frac{1}{3}$
4	1	.4138	.5177	.6836	.8025
	2	.0789	.1319	.2617	.4074
	3	.0071	.0162	.0508	.1111
	4	.0002	.0008	.0039	.0123
5	1	.4871	.5981	.7627	.8683
	2	.1207	.1962	.3672	.5391
	3	.0161	.0355	.1035	.2099
	4	.0011	.0033	.0156	.0453
	5		.0001	.0010	.0041
6	1	.5512	.6651	.8220	.9122
	2	.1665	.2632	.4661	.6488
	3	.0291	.0623	.1694	.3196
	4	.0030	.0087	.0376	.1001
	5	.0002	.0007	.0046	.0178
	6			.0002	.0014
7	1	.6073	.7209	.8665	.9415
	2	.2146	.3302	.5551	.7366
	3	.0463	.0958	.2436	.4294
	4	.0062	.0176	.0706	.1733
	5	.0005	.0020	.0129	.0453
	6		.0001	.0013	.0069
	7			.0001	.0005
8	1	.6564	.7674	.8999	.9610
	2	.2637	.3953	.6329	.8049
	3	.0673	.1348	.3215	.5318
	4	.0112	.0307	.1138	.2586
	5	.0012	.0046	.0273	.0879
	6	.0001	.0004	.0042	.0197
	7			.0004	.0026
	8				.0002
9	1	.6993	.8062	.9249	.9740
	2	.3128	.4573	.6997	.8569
	3	.0919	.1783	.3993	.6228
	4	.0183	.0480	.1657	.3497
	5	.0025	.0090	.0489	.1448
	6	.0002	.0011	.0100	.0424
	7		.0001	.0013	.0083
	8			.0001	.0010
	9				.0001
10	1	.7369	.8385	.9437	.9827
	2	.3611	.5155	.7560	.8959
	3	.1195	.2248	.4744	.7009
	4	.0275	.0697	.2241	.4407
	5	.0045	.0155	.0781	.2131
	6	.0005	.0024	.0197	.0766
	7		.0003	.0035	.0197
	8			.0004	.0034
	9				.0004
11	1	.7698	.8654	.9578	.9884
	2	.4081	.5693	.8029	.9249
	3	.1497	.2732	.5448	.7659
	4	.0390	.0956	.2867	.5274
	5	.0073	.0245	.1146	.2890
	6	.0010	.0046	.0343	.1221
	7	.0001	.0006	.0076	.0386
	8		.0001	.0012	.0088
	9			.0001	.0014
	10				.0001

n	r	$\pi=\frac{1}{8}$	$\frac{1}{6}$	$\frac{1}{4}$	$\frac{1}{3}$
12	1	.7986	.8878	.9683	.9923
	2	.4533	.6187	.8416	.9460
	3	.1820	.3226	.6093	.8189
	4	.0528	.1252	.3512	.6069
	5	.0113	.0363	.1576	.3685
	6	.0018	.0079	.0544	.1777
	7	.0002	.0013	.0143	.0664
	8		.0002	.0028	.0188
	9			.0004	.0039
	10				.0005
13	1	.8238	.9065	.9762	.9949
	2	.4965	.6635	.8733	.9615
	3	.2159	.3719	.6674	.8613
	4	.0690	.1581	.4157	.6776
	5	.0165	.0512	.2060	.4480
	6	.0030	.0127	.0802	.2413
	7	.0004	.0024	.0243	.1035
	8		.0003	.0056	.0346
	9			.0010	.0088
	10			.0001	.0016
	11				.0002
14	1	.8458	.9221	.9822	.9966
	2	.5374	.7040	.8990	.9726
	3	.2510	.4205	.7189	.8947
	4	.0873	.1937	.4787	.7388
	5	.0230	.0690	.2585	.5245
	6	.0047	.0191	.1117	.3102
	7	.0007	.0041	.0383	.1495
	8	.0001	.0007	.0103	.0576
	9		.0001	.0022	.0174
	10			.0003	.0040
	11				.0007
	12				.0001
15	1	.8651	.9351	.9866	.9977
	2	.5759	.7404	.9198	.9806
	3	.2868	.4678	.7639	.9206
	4	.1078	.2315	.5387	.7908
	5	.0311	.0898	.3135	.5959
	6	.0070	.0274	.1484	.3816
	7	.0012	.0066	.0566	.2030
	8	.0002	.0013	.0173	.0882
	9		.0002	.0042	.0308
	10			.0008	.0085
	11			.0001	.0018
	12				.0003
16	1	.8819	.9459	.9900	.9985
	2	.6121	.7728	.9365	.9863
	3	.3229	.5132	.8029	.9406
	4	.1302	.2709	.5950	.8341
	5	.0407	.1134	.3698	.6609
	6	.0100	.0378	.1897	.4531
	7	.0019	.0101	.0796	.2626
	8	.0003	.0021	.0271	.1265
	9		.0004	.0075	.0500
	10			.0016	.0159
	11			.0003	.0040
	12				.0008
	13				.0001

n	r	$\pi=\frac{1}{8}$	$\frac{1}{6}$	$\frac{1}{4}$	$\frac{1}{3}$
17	1	.8967	.9549	.9925	.9990
	2	.6458	.8017	.9499	.9904
	3	.3591	.5565	.8363	.9558
	4	.1543	.3113	.6470	.8696
	5	.0518	.1396	.4261	.7186
	6	.0138	.0504	.2347	.5223
	7	.0029	.0147	.1071	.3261
	8	.0005	.0035	.0402	.1719
	9	.0001	.0007	.0124	.0755
	10		.0001	.0031	.0273
	11			.0006	.0080
	12			.0001	.0019
	13				.0003
18	1	.9096	.9624	.9944	.9993
	2	.6772	.8272	.9605	.9932
	3	.3949	.5973	.8647	.9674
	4	.1799	.3521	.6943	.8983
	5	.0646	.1682	.4813	.7689
	6	.0186	.0653	.2825	.5878
	7	.0043	.0206	.1390	.3915
	8	.0008	.0053	.0569	.2233
	9	.0001	.0011	.0193	.1076
	10		.0002	.0054	.0433
	11			.0012	.0144
	12			.0002	.0039
	13				.0009
	14				.0001
19	1	.9209	.9687	.9958	.9995
	2	.7062	.8498	.9690	.9953
	3	.4302	.6357	.8887	.9760
	4	.2067	.3930	.7369	.9213
	5	.0791	.1989	.5346	.8121
	6	.0243	.0824	.3322	.6481
	7	.0061	.0281	.1749	.4569
	8	.0012	.0079	.0775	.2793
	9	.0002	.0018	.0287	.1462
	10		.0004	.0089	.0648
	11		.0001	.0023	.0241
	12			.0005	.0074
	13			.0001	.0019
	14				.0004
	15				.0001
20	1	.9308	.9739	.9968	.9997
	2	.7331	.8696	.9757	.9967
	3	.4647	.6713	.9087	.9824
	4	.2347	.4335	.7748	.9396
	5	.0950	.2313	.5852	.8485
	6	.0312	.1018	.3828	.7028
	7	.0084	.0371	.2142	.5207
	8	.0019	.0113	.1018	.3385
	9	.0003	.0028	.0409	.1905
	10	.0001	.0006	.0139	.0919
	11		.0001	.0039	.0376
	12			.0009	.0130
	13			.0002	.0037
	14				.0009
	15				.0002

Approximations to binomial distributions. For $n > 20$ two approximations to the distribution $\text{Bi}(n, \pi)$ are available. The Poisson approximation is preferable if $\pi < 0.25$ or > 0.75 and $n\pi(1-\pi) < 30$. If $0.25 < \pi < 0.75$ or $n\pi(1-\pi) > 30$ the normal approximation is preferable.

(i) *The Poisson approximation.* If s has the Poisson distribution $\text{Po}(n\pi - n\pi^2)$, the distribution of $r = s + n\pi^2$ approximates to the binomial distribution $\text{Bi}(n, \pi)$, which has the same mean and the same variance. The tail probability $Q(r)$ of $\text{Bi}(n, \pi)$ is therefore given approximately by the tail probability $Q(s)$ of $\text{Po}(n\pi - n\pi^2)$.

Example. If $r \sim \text{Bi}(50, 0.2)$ find $\text{Prob}(r \geqslant 15)$. Here $n\pi = 10$, $n\pi^2 = 2$, and so the required probability is given approximately by $Q(13)$ for $\text{Po}(8)$. The distribution of $\text{Po}(8)$ is given on p. 27; we find $Q(13) = 0.0638 \approx 0.064$. The true value is 0.0607.

If $n\pi^2$ lies between consecutive integers i and k the tail probability should be evaluated by interpolation between $Q(r-i)$ for $\text{Po}(n\pi - i)$ and $Q(r-k)$ for $\text{Po}(n\pi - k)$.

Example. The distribution of r is $\text{Bi}(50, 0.1)$; find $\text{Prob}(r \geqslant 4)$. Here $n\pi = 5$, and $n\pi^2 = 0.5$ which lies between 0 and 1. The required probability is the mean of $Q(4)$ for $\text{Po}(5)$ and $Q(3)$ for $\text{Po}(4)$; its value is 0.748. The true probability is 0.7497.

(ii) *The normal approximation.* The distribution $\text{Bi}(n, \pi)$ approximates to the normal distribution $\text{N}(n\pi, n\pi - n\pi^2)$. Consequently, if $r \sim \text{Bi}(n, \pi)$, $(r - n\pi)/\sqrt{(n\pi - n\pi^2)} \sim \text{N}(0, 1)$ approximately. $Q(r)$ is given approximately by $Q(z)$ for $\text{N}(0, 1)$ with $z = (r - n\pi - \frac{1}{2})/\sqrt{(n\pi - n\pi^2)}$. The $\frac{1}{2}$ in this expression is the usual correction for continuity.

Example. If $r \sim \text{Bi}(25, 0.36)$, find $\text{Prob}(r \geqslant 11)$. The required probability is given approximately by $Q(z)$ for the standard normal distribution with $z = (11 - 9 - \frac{1}{2})/2.4 = 0.625$. The table on p. 31 gives the value $0.2660 \approx 0.266$; the true value is 0.2624.

THE INVERSE SINE TRANSFORMATION

p	0	1	2	3	4	5	6	7	8	9	1	2	3	4	5	6	7	8	9
													Add proportional parts						
0.0	0.000	0.100	0.142	0.174	0.201	0.226	0.247												
								0.268	0.287	0.305	2	4	5	7	9	11	13	14	16
0.1	0.322	0.338	0.354	0.369	0.383						2	3	5	6	8	9	11	12	14
						0.398	0.412	0.425	0.438	0.451	1	3	4	5	7	8	9	10	12
0.2	0.464	0.476	0.488	0.500	0.512	0.524	0.535	0.546	0.558	0.569	1	2	3	5	6	7	8	9	10
0.3	0.580	0.591	0.601	0.612	0.623	0.633	0.644	0.654	0.664	0.674	1	2	3	4	5	6	7	8	9
0.4	0.685	0.695	0.705	0.715	0.725	0.735	0.745	0.755	0.765	0.775	1	2	3	4	5	6	7	8	9
0.5	0.785	0.795	0.805	0.815	0.825	0.835	0.846	0.856	0.866	0.876	1	2	3	4	5	6	7	8	9
0.6	0.886	0.896	0.907	0.917	0.927	0.938	0.948	0.959	0.970	0.980	1	2	3	4	5	6	7	8	9
0.7	0.991	1.002	1.013	1.024	1.036	1.047	1.059	1.071	1.083	1.095	1	2	3	5	6	7	8	9	10
0.8	1.107	1.120	1.133	1.146	1.159	1.173					1	3	4	5	7	8	9	11	12
							1.187	1.202	1.217	1.233	2	3	5	6	8	9	11	12	14
0.9	1.249	1.266	1.284								2	4	5	7	9	11	13	14	16
				1.303	1.323	1.345	1.369	1.397	1.429	1.471									

This table gives $\sin^{-1}\sqrt{p}$ expressed in radians. Interpolation is reliable for $0.02 < p < 0.98$.

If the distribution of r is $\text{Bi}(n, \pi)$, the variance of the proportion $p = r/n$ is $\pi(1-\pi)/n$, and thus varies with π. If $\theta = \sin^{-1}\sqrt{p}$ the variance of θ is approximately $1/(4n)$, which is independent of π. The transformation is useful when using methods that require homogeneity of variance (e.g. in an analysis of variance on proportions).

LOWER QUANTILES OF BINOMIAL DISTRIBUTIONS $\mathrm{Bi}(n, \tfrac{1}{2})$

n	.005	.01	.025	.05	.10	.25	n	.005	.01	.025	.05	.10	.25
6	0	0	1	1	1	2	52	17	18	19	20	21	24
7	0	1	1	1	2	3	54	18	19	20	21	22	25
8	1	1	1	2	2	3	56	18	19	21	22	23	25
9	1	1	2	2	3	3	58	19	20	22	23	24	26
10	1	1	2	2	3	4	60	20	21	22	24	25	27
11	1	2	2	3	3	4	62	21	22	23	25	26	28
12	2	2	3	3	4	5	64	22	23	24	25	27	29
13	2	2	3	4	4	5	66	23	24	25	26	28	30
14	2	3	3	4	5	6	68	23	24	26	27	29	31
15	3	3	4	4	5	6	70	24	25	27	28	30	32
16	3	3	4	5	5	7	72	25	26	28	29	31	33
17	3	4	5	5	6	7	74	26	27	29	30	31	34
18	4	4	5	6	6	8	76	27	28	29	31	32	35
19	4	5	5	6	7	8	78	28	29	30	32	33	36
20	4	5	6	6	7	8	80	29	30	31	33	34	37
21	5	5	6	7	8	9	82	29	31	32	34	35	38
22	5	6	6	7	8	9	84	30	31	33	34	36	39
23	5	6	7	8	8	10	86	31	32	34	35	37	40
24	6	6	7	8	9	10	88	32	33	35	36	38	41
25	6	7	8	8	9	11	90	33	34	36	37	39	42
26	7	7	8	9	10	11	92	34	35	37	38	40	43
27	7	8	8	9	10	12	94	35	36	38	39	41	44
28	7	8	9	10	11	12	96	35	37	38	40	42	45
29	8	8	9	10	11	13	98	36	38	39	41	43	46
30	8	9	10	11	11	13	100	37	38	40	42	44	47
31	8	9	10	11	12	14	110	42	43	45	46	48	51
32	9	9	10	11	12	14	120	46	47	49	51	53	56
33	9	10	11	12	13	15	130	50	52	54	56	58	61
34	10	10	11	12	13	15	140	55	56	58	60	62	66
35	10	11	12	13	14	16	150	59	61	63	65	67	71
36	10	11	12	13	14	16	160	64	65	68	70	72	76
37	11	11	13	14	15	16	170	68	70	72	74	77	81
38	11	12	13	14	15	17	180	73	74	77	79	81	85
39	12	12	13	14	16	17	190	77	79	82	84	86	90
40	12	13	14	15	16	18	200	82	84	86	88	91	95
41	12	13	14	15	16	18	220	91	93	95	98	100	105
42	13	14	15	16	17	19	240	100	102	105	107	110	115
43	13	14	15	16	17	19	260	109	111	114	117	120	125
44	14	14	16	17	18	20	280	118	121	124	126	129	134
45	14	15	16	17	18	20	300	128	130	133	136	139	144
46	14	15	16	17	19	21	320	137	139	142	145	149	154
47	15	16	17	18	19	21	340	146	149	152	155	158	164
48	15	16	17	18	20	22	360	156	158	161	164	168	174
49	16	16	18	19	20	22	380	165	167	171	174	178	183
50	16	17	18	19	20	23	400	174	177	180	184	187	193

Linear interpolation is permissible; the error is at most 1. Alternatively the normal approximation $N(\tfrac{1}{2}n, \tfrac{1}{4}n)$ can be used; fig. 7 compares the distributions $\mathrm{Bi}(12, \tfrac{1}{2})$ and $N(6, 3)$.

Upper quantiles are given by:

$$r_{[P]} = n - r_{[1-P]}$$

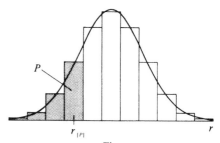

Fig. 7

24

The sign test. (a) Given a sample of matched pairs (x_i, y_i), $i = 1, 2, \ldots, n'$, taken from a bivariate population, the test is used to test the null hypothesis that $\text{Prob}(x < y) = \text{Prob}(x > y)$ for all pairs against the alternatives that $\text{Prob}(x < y) > \text{Prob}(x > y)$, or $\text{Prob}(x < y) < \text{Prob}(x > y)$ (one-sided tests), or that $\text{Prob}(x < y) \neq \text{Prob}(x > y)$ (two-sided test). We discard pairs in the sample with $y_i - x_i = 0$, leaving say r pairs with $y_i - x_i > 0$ and say $n - r$ pairs with $y_i - x_i < 0$ (so that $n \leqslant n'$). We take r as test statistic. The null distribution of r is $\text{Bi}(n, \tfrac{1}{2})$; this we test by reference to the table, using a one-tail or a two-tail test, as appropriate.

(b) As a particular case, given a sample $x_1, x_2, \ldots, x_n$ from a univariate population, we can test the null hypothesis that the population has a prescribed median M, i.e. $\text{Prob}(x < M) = \text{Prob}(x > M)$. We have merely to replace y_i by M.

(c) If the X- and Y-populations differ only in location, or if they are symmetric about their medians, test (a) becomes a test for difference between medians; i.e. we test the null hypothesis $M_X = M_Y$ against the alternatives $M_X < M_Y$, or $M_X > M_Y$, or $M_X \neq M_Y$.

The McNemar test. Consider a bivariate population of pairs (x, y), and suppose that the xs and ys are of two types, $+$ and $-$. Given a random sample of N pairs, we wish to test the null hypothesis that the proportions of positive elements in the X- and Y-populations are equal. The pairs are of four types, $++$, $+-$, $-+$ and $--$. If we set out their frequencies in a 2×2 array, the row total R_+ and the column total C_+ are the numbers of positive X-elements and Y-elements respectively. The null hypothesis requires that they do not differ significantly, and consequently that the frequencies f_{+-} and f_{-+} do not differ significantly. We take f_{+-} as test statistic, and take its

f_{++}	f_{+-}	R_+
f_{-+}	f_{--}	R_-
C_+	C_-	N

distribution to be $\text{Bi}(n, \tfrac{1}{2})$ with $n = f_{+-} + f_{-+}$. We test whether f_{+-} differs significantly from $\tfrac{1}{2}(f_{+-} + f_{-+})$ by use of the table.

Alternatively (unless $f_{+-} + f_{-+}$ is small) we use as test statistic

$$U = \frac{(f_{+-} - f_{-+})^2}{f_{+-} + f_{-+}}$$

The distribution of U is approximately $\chi^2(1)$ (see p. 38). With a correction for continuity the test statistic becomes:

$$U_{\text{corr}} = \frac{(|f_{+-} - f_{-+}| - 1)^2}{f_{+-} + f_{-+}}$$

Note that the upper tail of the chi-square distribution corresponds to both tails of the binomial distribution, so gives a two-sided test. The chi-square test should not be used unless both f_{+-} and f_{-+} are at least 5.

In some applications the xs are test scores of N individuals and the ys are scores of the same individuals in a repetition of the test after a conditioning event has occurred. In other applications the xs and ys are scores of matched pairs of individuals in a single test.

The Cox & Stuart test for trend. Given a sequence of random observations $x_1, x_2, \ldots, x_m$, we wish to test for the presence of a trend. We consider the pairs (x_1, x_{1+c}), (x_2, x_{2+c}), $\ldots, (x_{m-c}, x_m)$, where c is either $m/2$ or $(m+1)/2$. We treat these pairs as in the sign test (a). A significant predominance of positive or negative pairs indicates an upward or downward trend.

The observations may be arranged in the order in which they were made. Or they may be arranged in the order of magnitude of their predicted values; the test is then a test of correlation between the observations and their predicted values.

It is sometimes advantageous to use fewer pairs, more widely separated. For example, for $m = 15$ take $c = 10$.

UPPER TAIL PROBABILITIES $Q(r)$ OF POISSON DISTRIBUTIONS Po(λ)

r	0.02	0.04	0.06	0.08	0.1	0.2	λ 0.3	0.4	0.5	0.6	0.7	0.8	0.9
1	.0198	.0392	.0582	.0769	.0952	.1813	.2592	.3297	.3935	.4512	.5034	.5507	.5934
2	.0002	.0008	.0017	.0030	.0047	.0175	.0369	.0616	.0902	.1219	.1558	.1912	.2275
3				.0001	.0002	.0011	.0036	.0079	.0144	.0231	.0341	.0474	.0629
4						.0001	.0003	.0008	.0018	.0034	.0058	.0091	.0135
5								.0001	.0002	.0004	.0008	.0014	.0023
6											.0001	.0002	.0003

r	1.0	1.1	1.2	1.3	1.4	1.5	λ 1.6	1.7	1.8	1.9	2.0	2.1	2.2
1	.6321	.6671	.6988	.7275	.7534	.7769	.7981	.8173	.8347	.8504	.8647	.8775	.8892
2	.2642	.3010	.3374	.3732	.4082	.4422	.4751	.5068	.5372	.5663	.5940	.6204	.6454
3	.0803	.0996	.1205	.1429	.1665	.1912	.2166	.2428	.2694	.2963	.3233	.3504	.3773
4	.0190	.0257	.0338	.0431	.0537	.0656	.0788	.0932	.1087	.1253	.1429	.1614	.1806
5	.0037	.0054	.0077	.0107	.0143	.0186	.0237	.0296	.0364	.0441	.0527	.0621	.0725
6	.0006	.0010	.0015	.0022	.0032	.0045	.0060	.0080	.0104	.0132	.0166	.0204	.0249
7	.0001	.0001	.0003	.0004	.0006	.0009	.0013	.0019	.0026	.0034	.0045	.0059	.0075
8				.0001	.0001	.0002	.0003	.0004	.0006	.0008	.0011	.0015	.0020
9								.0001	.0001	.0002	.0002	.0003	.0005
10												.0001	.0001

r	2.4	2.6	2.8	3.0	3.2	3.4	λ 3.6	3.8	4.0	4.2	4.4	4.6	4.8
1	.9093	.9257	.9392	.9502	.9592	.9666	.9727	.9776	.9817	.9850	.9877	.9899	.9918
2	.6916	.7326	.7689	.8009	.8288	.8532	.8743	.8926	.9084	.9220	.9337	.9437	.9523
3	.4303	.4816	.5305	.5768	.6201	.6603	.6973	.7311	.7619	.7898	.8149	.8374	.8575
4	.2213	.2640	.3081	.3528	.3975	.4416	.4848	.5265	.5665	.6046	.6406	.6743	.7058
5	.0959	.1226	.1523	.1847	.2194	.2558	.2936	.3322	.3712	.4102	.4488	.4868	.5237
6	.0357	.0490	.0651	.0839	.1054	.1295	.1559	.1844	.2149	.2469	.2801	.3142	.3490
7	.0116	.0172	.0244	.0335	.0446	.0579	.0733	.0909	.1107	.1325	.1564	.1820	.2092
8	.0033	.0053	.0081	.0119	.0168	.0231	.0308	.0401	.0511	.0639	.0786	.0951	.1133
9	.0009	.0015	.0024	.0038	.0057	.0083	.0117	.0160	.0214	.0279	.0358	.0451	.0558
10	.0002	.0004	.0007	.0011	.0018	.0027	.0040	.0058	.0081	.0111	.0149	.0195	.0251
11		.0001	.0002	.0003	.0005	.0008	.0013	.0019	.0028	.0041	.0057	.0078	.0104
12				.0001	.0001	.0002	.0004	.0006	.0009	.0014	.0020	.0029	.0040
13						.0001	.0001	.0002	.0003	.0004	.0007	.0010	.0014
14									.0001	.0001	.0002	.0003	.0005
15											.0001	.0001	.0001

If, in a Poisson process, the expected number of events occurring in an interval of length τ is λ, the probability that exactly r events occur in that interval is given by the Poisson distribution Po(λ) (or P(λ)):

$$p(r) = e^{-\lambda}\lambda^r/r!$$

The mean and variance of the distribution are both equal to λ.

The table gives, for $r > 0$, the probability $Q(r)$ of r or more events: $Q(r) = \Sigma_{x=r}^{\infty} p(x)$. $Q(0)$ always has the value 1 and is not tabulated. An entry 1− indicates a probability between 1 and 0.99995. Where a space is left blank the probability is less than 0.00005. The individual terms can be found by subtraction:

$$p(r) = Q(r) - Q(r+1)$$

The probability $P(r)$ of r or fewer events (*the cumulative probability*) is given by:

$$P(r) = 1 - Q(r+1)$$

UPPER TAIL PROBABILITIES $Q(r)$ OF POISSON DISTRIBUTIONS $\text{Po}(\lambda)$

r	5.0	5.2	5.4	5.6	5.8	6.0	6.2	6.4	6.6	6.8	7.0	7.2	7.4
1	.9933	.9945	.9955	.9963	.9970	.9975	.9980	.9983	.9986	.9989	.9991	.9993	.9994
2	.9596	.9658	.9711	.9756	.9794	.9826	.9854	.9877	.9897	.9913	.9927	.9939	.9949
3	.8753	.8912	.9052	.9176	.9285	.9380	.9464	.9537	.9600	.9656	.9704	.9745	.9781
4	.7350	.7619	.7867	.8094	.8300	.8488	.8658	.8811	.8948	.9072	.9182	.9281	.9368
5	.5595	.5939	.6267	.6579	.6873	.7149	.7408	.7649	.7873	.8080	.8270	.8445	.8605
6	.3840	.4191	.4539	.4881	.5217	.5543	.5859	.6163	.6453	.6730	.6993	.7241	.7474
7	.2378	.2676	.2983	.3297	.3616	.3937	.4258	.4577	.4892	.5201	.5503	.5796	.6080
8	.1334	.1551	.1783	.2030	.2290	.2560	.2840	.3127	.3419	.3715	.4013	.4311	.4607
9	.0681	.0819	.0974	.1143	.1328	.1528	.1741	.1967	.2204	.2452	.2709	.2973	.3243
10	.0318	.0397	.0488	.0591	.0708	.0839	.0984	.1142	.1314	.1498	.1695	.1904	.2123
11	.0137	.0177	.0225	.0282	.0349	.0426	.0514	.0614	.0726	.0849	.0985	.1133	.1293
12	.0055	.0073	.0096	.0125	.0160	.0201	.0250	.0307	.0373	.0448	.0534	.0629	.0735
13	.0020	.0028	.0038	.0051	.0068	.0088	.0113	.0143	.0179	.0221	.0270	.0327	.0391
14	.0007	.0010	.0014	.0020	.0027	.0036	.0048	.0063	.0080	.0102	.0128	.0159	.0195
15	.0002	.0003	.0005	.0007	.0010	.0014	.0019	.0026	.0034	.0044	.0057	.0073	.0092
16	.0001	.0001	.0002	.0002	.0004	.0005	.0007	.0010	.0014	.0018	.0024	.0031	.0041
17			.0001	.0001	.0001	.0002	.0003	.0004	.0005	.0007	.0010	.0013	.0017
18						.0001	.0001	.0001	.0002	.0003	.0004	.0005	.0007
19									.0001	.0001	.0001	.0002	.0003
20												.0001	.0001

r	7.6	7.8	8.0	8.2	8.4	8.6	8.8	9.0	9.2	9.4	9.6	9.8	10.0
1	.9995	.9996	.9997	.9997	.9998	.9998	.9998	.9999	.9999	.9999	.9999	.9999	1 −
2	.9957	.9964	.9970	.9975	.9979	.9982	.9985	.9988	.9990	.9991	.9993	.9994	.9995
3	.9812	.9839	.9862	.9882	.9900	.9914	.9927	.9938	.9947	.9955	.9962	.9967	.9972
4	.9446	.9515	.9576	.9630	.9677	.9719	.9756	.9788	.9816	.9840	.9862	.9880	.9897
5	.8751	.8883	.9004	.9113	.9211	.9299	.9379	.9450	.9514	.9571	.9622	.9667	.9707
6	.7693	.7897	.8088	.8264	.8427	.8578	.8716	.8843	.8959	.9065	.9162	.9250	.9329
7	.6354	.6616	.6866	.7104	.7330	.7543	.7744	.7932	.8108	.8273	.8426	.8567	.8699
8	.4900	.5188	.5470	.5746	.6013	.6272	.6522	.6761	.6990	.7208	.7416	.7612	.7798
9	.3518	.3796	.4075	.4353	.4631	.4906	.5177	.5443	.5704	.5958	.6204	.6442	.6672
10	.2351	.2589	.2834	.3085	.3341	.3600	.3863	.4126	.4389	.4651	.4911	.5168	.5421
11	.1465	.1648	.1841	.2045	.2257	.2478	.2706	.2940	.3180	.3424	.3671	.3920	.4170
12	.0852	.0980	.1119	.1269	.1429	.1600	.1780	.1970	.2168	.2374	.2588	.2807	.3032
13	.0464	.0546	.0638	.0739	.0850	.0971	.1102	.1242	.1393	.1552	.1721	.1899	.2084
14	.0238	.0286	.0342	.0405	.0476	.0555	.0642	.0739	.0844	.0958	.1081	.1214	.1355
15	.0114	.0141	.0173	.0209	.0251	.0299	.0353	.0415	.0483	.0559	.0643	.0735	.0835
16	.0052	.0066	.0082	.0102	.0125	.0152	.0184	.0220	.0262	.0309	.0362	.0421	.0487
17	.0022	.0029	.0037	.0047	.0059	.0074	.0091	.0111	.0135	.0162	.0194	.0230	.0270
18	.0009	.0012	.0016	.0021	.0027	.0034	.0043	.0053	.0066	.0081	.0098	.0119	.0143
19	.0004	.0005	.0006	.0009	.0011	.0015	.0019	.0024	.0031	.0038	.0048	.0059	.0072
20	.0001	.0002	.0003	.0003	.0005	.0006	.0008	.0011	.0014	.0017	.0022	.0028	.0035
21		.0001	.0001	.0001	.0002	.0002	.0003	.0004	.0006	.0008	.0010	.0012	.0016
22					.0001	.0001	.0001	.0002	.0002	.0003	.0004	.0005	.0007
23								.0001	.0001	.0001	.0002	.0002	.0003
24											.0001	.0001	.0001

Linear interpolation with respect to λ may lead to an error of a unit or more in the third decimal place.

UPPER TAIL PROBABILITIES $Q(r)$ OF POISSON DISTRIBUTIONS $Po(\lambda)$

r	λ 10.0	10.5	11.0	11.5	12.0	12.5	13.0	13.5	14.0	14.5	15.0	16.0	17.0
1	1 −	1 −	1 −	1 −	1 −	1 −							
2	.9995	.9997	.9998	.9999	.9999	.9999	1 −	1 −	1 −	1 −			
3	.9972	.9982	.9988	.9992	.9995	.9997	.9998	.9999	.9999	.9999	1 −	1 −	
4	.9897	.9929	.9951	.9966	.9977	.9984	.9990	.9993	.9995	.9997	.9998	.9999	1 −
5	.9707	.9789	.9849	.9893	.9924	.9947	.9963	.9974	.9982	.9988	.9991	.9996	.9998
6	.9329	.9496	.9625	.9723	.9797	.9852	.9893	.9923	.9945	.9961	.9972	.9986	.9993
7	.8699	.8984	.9214	.9397	.9542	.9654	.9741	.9807	.9858	.9895	.9924	.9960	.9979
8	.7798	.8215	.8568	.8863	.9105	.9302	.9460	.9585	.9684	.9761	.9820	.9900	.9946
9	.6672	.7206	.7680	.8094	.8450	.8751	.9002	.9210	.9379	.9516	.9626	.9780	.9874
10	.5421	.6029	.6595	.7112	.7576	.7986	.8342	.8647	.8906	.9122	.9301	.9567	.9739
11	.4170	.4793	.5401	.5983	.6528	.7029	.7483	.7888	.8243	.8551	.8815	.9226	.9509
12	.3032	.3613	.4207	.4802	.5384	.5942	.6468	.6955	.7400	.7799	.8152	.8730	.9153
13	.2084	.2580	.3113	.3671	.4240	.4810	.5369	.5907	.6415	.6889	.7324	.8069	.8650
14	.1355	.1747	.2187	.2670	.3185	.3722	.4270	.4818	.5356	.5875	.6368	.7255	.7991
15	.0835	.1121	.1460	.1847	.2280	.2750	.3249	.3767	.4296	.4824	.5343	.6325	.7192
16	.0487	.0683	.0926	.1217	.1556	.1940	.2364	.2822	.3306	.3808	.4319	.5333	.6285
17	.0270	.0396	.0559	.0764	.1013	.1307	.1645	.2025	.2441	.2888	.3359	.4340	.5323
18	.0143	.0219	.0322	.0458	.0630	.0842	.1095	.1391	.1728	.2103	.2511	.3407	.4360
19	.0072	.0115	.0177	.0262	.0374	.0519	.0698	.0916	.1174	.1470	.1805	.2577	.3450
20	.0035	.0058	.0093	.0143	.0213	.0306	.0427	.0579	.0765	.0988	.1248	.1878	.2637
21	.0016	.0028	.0047	.0075	.0116	.0173	.0250	.0351	.0479	.0638	.0830	.1318	.1945
22	.0007	.0013	.0023	.0038	.0061	.0094	.0141	.0204	.0288	.0396	.0531	.0892	.1385
23	.0003	.0006	.0010	.0018	.0030	.0049	.0076	.0115	.0167	.0237	.0327	.0582	.0953
24	.0001	.0002	.0005	.0008	.0015	.0025	.0040	.0062	.0093	.0137	.0195	.0367	.0633
25		.0001	.0002	.0004	.0007	.0012	.0020	.0032	.0050	.0076	.0112	.0223	.0406
26			.0001	.0002	.0003	.0006	.0010	.0016	.0026	.0041	.0062	.0131	.0252
27				.0001	.0001	.0003	.0005	.0008	.0013	.0021	.0033	.0075	.0152
28					.0001	.0001	.0002	.0004	.0006	.0011	.0017	.0041	.0088
29						.0001	.0001	.0002	.0003	.0005	.0009	.0022	.0050
30								.0001	.0001	.0002	.0004	.0011	.0027
31									.0001	.0001	.0002	.0006	.0014
32											.0001	.0003	.0007
33												.0001	.0004
34												.0001	.0002
35													.0001

Normal approximation. For $\lambda > 30$ a Poisson distribution approximates to the normal distribution with the same mean and the same variance; the tail probability $Q(r)$ for $Po(\lambda)$ is approximately equal to the corresponding tail probability $Q(z)$ for N(0, 1) with $z = (r - \lambda - \frac{1}{2})/\sqrt{\lambda}$, the $\frac{1}{2}$ in this expression being the usual correction for continuity. Errors of several units in the third decimal place may occur.

Example. Approximate values of $Q(30)$, $Q(40)$ and $Q(50)$ for the distribution $Po(36)$ are given by $Q(z)$ for N(0, 1) with z equal to $-6.5/6$, $3.5/6$, and $13.5/6$ respectively; they are 0.861, 0.280 and 0.012. The true values are 0.8621, 0.2737 and 0.0156.

The relation between the Poisson and chi-square distributions. The tail probabilities are related. For even ν:

$$\text{Prob}\,(W \leqslant w \mid W \sim \chi^2(\nu)) = \text{Prob}\,(R \geqslant \tfrac{1}{2}\nu \mid R \sim Po(\tfrac{1}{2}w))$$

Example. If $W \sim \chi^2(18)$, $\text{Prob}\,(W \leqslant 26)$ is given by the upper tail probability $Q(9)$ of $Po(13)$, which has the value 0.9002 (see above). For comparison the quantile $w_{[.9]}$ of $\chi^2(18)$ has the value 25.99 (see p. 38).

UPPER TAIL PROBABILITIES $Q(r)$ OF POISSON DISTRIBUTIONS Po(λ)

r	18.0	19.0	20.0	21.0	22.0	23.0	24.0	25.0	26.0	27.0	28.0	29.0	30.0
4	I —												
5	.9999	I —	I —										
6	.9997	.9998	.9999	I —	I —								
7	.9990	.9995	.9997	.9999	.9999	I —							
8	.9971	.9985	.9992	.9996	.9998	.9999	I —	I —					
9	.9929	.9961	.9979	.9989	.9994	.9997	.9998	.9999	I —	I —			
10	.9846	.9911	.9950	.9972	.9985	.9992	.9996	.9998	.9999	.9999	I —		
11	.9696	.9817	.9892	.9937	.9965	.9980	.9989	.9994	.9997	.9998	.9999	I —	I —
12	.9451	.9653	.9786	.9871	.9924	.9956	.9975	.9986	.9992	.9996	.9998	.9999	.9999
13	.9083	.9394	.9610	.9755	.9849	.9909	.9946	.9969	.9982	.9990	.9994	.9997	.9998
14	.8574	.9016	.9339	.9566	.9722	.9826	.9893	.9935	.9962	.9978	.9987	.9993	.9996
15	.7919	.8503	.8951	.9284	.9523	.9689	.9802	.9876	.9924	.9954	.9973	.9984	.9991
16	.7133	.7852	.8435	.8889	.9231	.9480	.9656	.9777	.9858	.9912	.9946	.9967	.9981
17	.6249	.7080	.7789	.8371	.8830	.9179	.9437	.9623	.9752	.9840	.9899	.9937	.9961
18	.5314	.6216	.7030	.7730	.8310	.8772	.9129	.9395	.9589	.9726	.9821	.9885	.9927
19	.4378	.5305	.6186	.6983	.7675	.8252	.8717	.9080	.9354	.9555	.9700	.9801	.9871
20	.3491	.4394	.5297	.6157	.6940	.7623	.8197	.8664	.9032	.9313	.9522	.9674	.9781
21	.2693	.3528	.4409	.5290	.6131	.6899	.7574	.8145	.8613	.8985	.9273	.9489	.9647
22	.2009	.2745	.3563	.4423	.5284	.6106	.6861	.7527	.8095	.8564	.8940	.9233	.9456
23	.1449	.2069	.2794	.3595	.4436	.5277	.6083	.6825	.7483	.8048	.8517	.8896	.9194
24	.1011	.1510	.2125	.2840	.3626	.4449	.5272	.6061	.6791	.7441	.8002	.8471	.8854
25	.0683	.1067	.1568	.2178	.2883	.3654	.4460	.5266	.6041	.6758	.7401	.7958	.8428
26	.0446	.0731	.1122	.1623	.2229	.2923	.3681	.4471	.5261	.6021	.6728	.7363	.7916
27	.0282	.0486	.0779	.1174	.1676	.2277	.2962	.3706	.4481	.5256	.6003	.6699	.7327
28	.0173	.0313	.0525	.0825	.1225	.1726	.2323	.2998	.3730	.4491	.5251	.5986	.6671
29	.0103	.0195	.0343	.0564	.0871	.1274	.1775	.2366	.3033	.3753	.4500	.5247	.5969
30	.0059	.0118	.0218	.0374	.0602	.0915	.1321	.1821	.2407	.3065	.3774	.4508	.5243
31	.0033	.0070	.0135	.0242	.0405	.0640	.0958	.1367	.1866	.2447	.3097	.3794	.4516
32	.0018	.0040	.0081	.0152	.0265	.0436	.0678	.1001	.1411	.1908	.2485	.3126	.3814
33	.0010	.0022	.0047	.0093	.0169	.0289	.0467	.0715	.1042	.1454	.1949	.2521	.3155
34	.0005	.0012	.0027	.0055	.0105	.0187	.0314	.0498	.0751	.1082	.1495	.1989	.2556
35	.0002	.0006	.0015	.0032	.0064	.0118	.0206	.0338	.0528	.0787	.1121	.1535	.2027
36	.0001	.0003	.0008	.0018	.0038	.0073	.0132	.0225	.0363	.0559	.0822	.1159	.1574
37	.0001	.0002	.0004	.0010	.0022	.0044	.0082	.0146	.0244	.0388	.0589	.0856	.1196
38		.0001	.0002	.0005	.0012	.0026	.0050	.0092	.0160	.0263	.0413	.0619	.0890
39			.0001	.0003	.0007	.0015	.0030	.0057	.0103	.0175	.0283	.0438	.0648
40			.0001	.0001	.0004	.0008	.0017	.0034	.0064	.0113	.0190	.0303	.0463
41				.0001	.0002	.0004	.0010	.0020	.0039	.0072	.0125	.0205	.0323
42					.0001	.0002	.0005	.0012	.0024	.0045	.0080	.0136	.0221
43						.0001	.0003	.0007	.0014	.0027	.0050	.0089	.0148
44						.0001	.0002	.0004	.0008	.0016	.0031	.0056	.0097
45							.0001	.0002	.0004	.0009	.0019	.0035	.0063
46								.0001	.0002	.0005	.0011	.0022	.0040
47								.0001	.0001	.0003	.0006	.0013	.0025
48									.0001	.0002	.0004	.0008	.0015
49										.0001	.0002	.0004	.0009
50											.0001	.0002	.0005
51											.0001	.0001	.0003
52												.0001	.0002
53													.0001
54													.0001

Source: For the tables on pp. 26–9, *Poisson's Exponential Binomial Limit* (E. C. Molina).

PROBABILITY DENSITY $\phi(z)$ OF THE NORMAL DISTRIBUTION N(0,1)

z	0	1	2	3	4	5	6	7	8	9
0.	0.399	.397	.391	.381	.368	.352	.333	.312	.290	.266
1.	0.242	.218	.194	.171	.150	.130	.111	.094	.079	.066
2.	0.0540	.0440	.0355	.0283	.0224	.0175	.0136	.0104	.0079	.0060
3.	0.00443	.00327	.00238	.00172	.00123	.00087	.00061	.00042	.00029	.00020
4.	0.0^3134	$.0^489$	$.0^459$	$.0^439$	$.0^425$	$.0^416$	$.0^410$	$.0^564$	$.0^540$	$.0^524$

For $z < 0$ use the relation:
$$\phi(z) = \phi(-z)$$

The tabulated function is defined thus:

$$\phi(z) = \sqrt{\left(\frac{1}{2\pi}\right)}\exp\left(-\tfrac{1}{2}z^2\right)$$

In fig. 8 the probability density is represented by the ordinate of the graph, and the tail probabilities are represented by areas under the graph.
  The probability density of $N(\mu, \sigma^2)$ is

$$f(x) = \frac{1}{\sigma}\phi(z)$$

with $z = (x-\mu)/\sigma$.

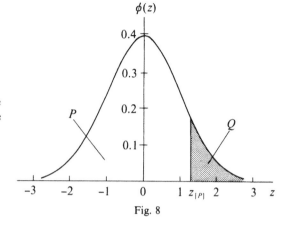

Fig. 8

UPPER QUANTILES $z_{[P]}$ OF THE NORMAL DISTRIBUTION N(0,1)

P	Q	z	P	Q	z	P	Q	z	P	Q	z	Q	z
.50	.50	0.000	.85	.15	1.036	.975	.025	1.960	.990	.010	2.326	$.0^34$	3.353
.55	.45	0.126	.86	.14	1.080	.976	.024	1.977	.991	.009	2.366	$.0^33$	3.432
.60	.40	0.253	.87	.13	1.126	.977	.023	1.995	.992	.008	2.409	$.0^32$	3.540
.65	.35	0.385	.88	.12	1.175	.978	.022	2.014	.993	.007	2.457	$.0^31$	3.719
.70	.30	0.524	.89	.11	1.227	.979	.021	2.034	.994	.006	2.512	$.0^45$	3.891
.75	.25	0.674	.90	.10	1.282	.980	.020	2.054	.995	.005	2.576	$.0^41$	4.265
.76	.24	0.706	.91	.09	1.341	.981	.019	2.075	.996	.004	2.652	$.0^55$	4.417
.77	.23	0.739	.92	.08	1.405	.982	.018	2.097	.997	.003	2.748	$.0^51$	4.753
.78	.22	0.772	.93	.07	1.476	.983	.017	2.120	.998	.002	2.878	$.0^65$	4.892
.79	.21	0.806	.94	.06	1.555	.984	.016	2.144	.999	.001	3.090	$.0^61$	5.199
.80	.20	0.842	.950	.050	1.645	.985	.015	2.170	.9991	$.0^39$	3.121	$.0^75$	5.327
.81	.19	0.878	.955	.045	1.695	.986	.014	2.197	.9992	$.0^38$	3.156	$.0^71$	5.612
.82	.18	0.915	.960	.040	1.751	.987	.013	2.226	.9993	$.0^37$	3.195	$.0^85$	5.731
.83	.17	0.954	.965	.035	1.812	.988	.012	2.257	.9994	$.0^36$	3.239	$.0^81$	5.998
.84	.16	0.994	.970	.030	1.881	.989	.011	2.290	.9995	$.0^35$	3.291	$.0^95$	6.109

The definition of $z_{[P]}$ is:

$$\int_{-\infty}^{z_{[P]}} \phi(u)\,du = P$$

If $Z \sim N(0, 1)$, $\mathrm{Prob}(Z < z_{[P]}) = P$, $\mathrm{Prob}(Z > z_{[P]}) = 1 - P = Q$, and (for $P > \tfrac{1}{2}$) $\mathrm{Prob}(|Z| > z_{[P]}) = 2Q$.
  Lower quantiles ($P < \tfrac{1}{2}$) are given by:

$$z_{[P]} = -z_{[1-P]}$$

UPPER TAIL PROBABILITIES $Q(z)$ OF THE NORMAL DISTRIBUTION N(0,1)

z	0	1	2	3	4	5	6	7	8	9	1 2 3 4 5 6 7 8 9 Subtract
0.0	.5000	.4960	.4920	.4880	.4840	.4801	.4761	.4721	.4681	.4641	4 8 12 16 20 24 28 32 36
0.1	.4602	.4562	.4522	.4483	.4443	.4404	.4364	.4325	.4286	.4247	4 8 12 16 20 24 28 32 36
0.2	.4207	.4168	.4129	.4090	.4052	.4013	.3974	.3936	.3897	.3859	4 8 12 15 19 23 27 31 35
0.3	.3821	.3783	.3745	.3707	.3669	.3632	.3594	.3557	.3520	.3483	4 7 11 15 19 22 26 30 34
0.4	.3446	.3409	.3372	.3336	.3300	.3264	.3228	.3192	.3156	.3121	4 7 11 14 18 22 25 29 32
0.5	.3085	.3050	.3015	.2981	.2946	.2912	.2877	.2843	.2810	.2776	3 7 10 14 17 20 24 27 31
0.6	.2743	.2709	.2676	.2643	.2611	.2578	.2546	.2514	.2483	.2451	3 7 10 13 16 19 23 26 29
0.7	.2420	.2389	.2358	.2327	.2296	.2266	.2236	.2206	.2177	.2148	3 6 9 12 15 18 21 24 27
0.8	.2119	.2090	.2061	.2033	.2005	.1977	.1949	.1922	.1894	.1867	3 5 8 11 14 16 19 22 25
0.9	.1841	.1814	.1788	.1762	.1736	.1711	.1685	.1660	.1635	.1611	3 5 8 10 13 15 18 20 23
1.0	.1587	.1562	.1539	.1515	.1492	.1469	.1446	.1423	.1401	.1379	2 5 7 9 12 14 16 19 21
1.1	.1357	.1335	.1314	.1292	.1271	.1251	.1230	.1210	.1190	.1170	2 4 6 8 10 12 14 16 18
1.2	.1151	.1131	.1112	.1093	.1075	.1056	.1038	.1020	.1003	.0985	2 4 6 7 9 11 13 15 17
1.3	.0968	.0951	.0934	.0918	.0901	.0885	.0869	.0853	.0838	.0823	2 3 5 6 8 10 11 13 14
1.4	.0808	.0793	.0778	.0764	.0749	.0735	.0721	.0708	.0694	.0681	1 3 4 6 7 8 10 11 13
1.5	.0668	.0655	.0643	.0630	.0618	.0606	.0594	.0582	.0571	.0559	1 2 4 5 6 7 8 10 11
1.6	.0548	.0537	.0526	.0516	.0505	.0495	.0485	.0475	.0465	.0455	1 2 3 4 5 6 7 8 9
1.7	.0446	.0436	.0427	.0418	.0409	.0401	.0392	.0384	.0375	.0367	1 2 3 4 4 5 6 7 8
1.8	.0359	.0351	.0344	.0336	.0329	.0322	.0314	.0307	.0301	.0294	1 1 2 3 4 4 5 6 6
1.9	.0287	.0281	.0274	.0268	.0262	.0256	.0250	.0244	.0239	.0233	1 1 2 2 3 4 4 5 5
2.0	.0228	.0222	.0217	.0212	.0207	.0202	.0197	.0192	.0188	.0183	0 1 1 2 2 3 3 4 4
2.1	.0179	.0174	.0170	.0166	.0162	.0158	.0154	.0150	.0146	.0143	0 1 1 2 2 2 3 3 4
2.2	.0139	.0136	.0132	.0129	.0125	.0122	.0119	.0116	.0113	.0110	0 1 1 1 2 2 2 3 3
2.3	.0107	.0104	.0102								0 1 1 1 1 2 2 2 2
				$.0^2990$	$.0^2964$	$.0^2939$	$.0^2914$				3 5 8 10 13 15 18 20 23
								$.0^2889$	$.0^2866$	$.0^2842$	2 5 7 9 12 14 16 18 21
2.4	$.0^2820$	$.0^2798$	$.0^2776$	$.0^2755$	$.0^2734$						2 4 6 8 11 13 15 17 19
						$.0^2714$	$.0^2695$	$.0^2676$	$.0^2657$	$.0^2639$	2 4 6 7 9 11 13 15 17
2.5	$.0^2621$	$.0^2604$	$.0^2587$	$.0^2570$	$.0^2554$	$.0^2539$	$.0^2523$	$.0^2508$	$.0^2494$	$.0^2480$	2 3 5 6 8 9 11 12 14
2.6	$.0^2466$	$.0^2453$	$.0^2440$	$.0^2427$	$.0^2415$	$.0^2402$	$.0^2391$	$.0^2379$	$.0^2368$	$.0^2357$	1 2 3 5 6 7 8 9 10
2.7	$.0^2347$	$.0^2336$	$.0^2326$	$.0^2317$	$.0^2307$	$.0^2298$	$.0^2289$	$.0^2280$	$.0^2272$	$.0^2264$	1 2 3 4 5 6 7 8 9
2.8	$.0^2256$	$.0^2248$	$.0^2240$	$.0^2233$	$.0^2226$	$.0^2219$	$.0^2212$	$.0^2205$	$.0^2199$	$.0^2193$	1 1 2 3 4 4 5 6 6
2.9	$.0^2187$	$.0^2181$	$.0^2175$	$.0^2169$	$.0^2164$	$.0^2159$	$.0^2154$	$.0^2149$	$.0^2144$	$.0^2139$	0 1 1 2 2 3 3 4 4
3.0	$.0^2135$	$.0^2131$	$.0^2126$	$.0^2122$	$.0^2118$	$.0^2114$	$.0^2111$	$.0^2107$	$.0^2104$	$.0^2100$	0 1 1 2 2 2 3 3 4
3.1	$.0^3968$	$.0^3935$	$.0^3904$								3 6 9 13 16 19 22 25 28
				$.0^3874$	$.0^3845$	$.0^3816$	$.0^3789$				3 6 8 11 14 17 20 22 25
								$.0^3762$	$.0^3736$	$.0^3711$	2 5 7 10 12 15 17 20 22
3.2	$.0^3687$	$.0^3664$	$.0^3641$	$.0^3619$	$.0^3598$						2 4 7 9 11 13 15 18 20
						$.0^3577$	$.0^3557$	$.0^3538$	$.0^3519$	$.0^3501$	2 4 6 8 9 11 13 15 17
3.3	$.0^3483$	$.0^3466$	$.0^3450$	$.0^3434$	$.0^3419$						2 3 5 6 8 10 11 13 14
						$.0^3404$	$.0^3390$	$.0^3376$	$.0^3362$	$.0^3349$	1 3 4 5 7 8 9 10 12
3.4	$.0^3337$	$.0^3325$	$.0^3313$	$.0^3302$	$.0^3291$	$.0^3280$	$.0^3270$	$.0^3260$	$.0^3251$	$.0^3242$	1 2 3 4 5 6 7 8 9
3.5	$.0^3233$	$.0^3224$	$.0^3216$	$.0^3208$	$.0^3200$	$.0^3193$	$.0^3185$	$.0^3178$	$.0^3172$	$.0^3165$	1 1 2 3 4 4 5 6 7
3.6	$.0^3159$	$.0^3153$	$.0^3147$	$.0^3142$	$.0^3136$	$.0^3131$	$.0^3126$	$.0^3121$	$.0^3117$	$.0^3112$	0 1 1 2 2 3 3 4 5
3.7	$.0^3108$	$.0^3104$	$.0^3100$	$.0^496$	$.0^492$	$.0^488$	$.0^485$	$.0^482$	$.0^478$	$.0^475$	
3.8	$.0^472$	$.0^469$	$.0^467$	$.0^464$	$.0^462$	$.0^459$	$.0^457$	$.0^454$	$.0^452$	$.0^450$	
3.9	$.0^448$	$.0^446$	$.0^444$	$.0^442$	$.0^441$	$.0^439$	$.0^437$	$.0^436$	$.0^434$	$.0^433$	

The definition of $Q(z)$ is:

$$Q(z) = \int_z^\infty \phi(u)\, du$$

For negative z use the relation: $\quad Q(z) = 1 - Q(-z)$

The approximation

$$-\lg Q \approx 0.3991 + \lg z + 0.4342945\{\tfrac{1}{2}z^2 + (z^2 + 2.25)^{-1}\}$$

may be used. The error is less than 0.001 for $z > 3.2$, and less than 0.0001 for $z > 4$.

If $V \sim N(\mu, \sigma^2)$, $\mathrm{Prob}(V > x)$ is given by $Q(z)$ with $z = (x - \mu)/\sigma$.

NORMAL SCORES (RANKITS) $E(n, k)$

k	1	2	3	4	5	n 6	7	8	9	10
1	0.0000	0.5642	0.8463	1.0294	1.1630	1.2672	1.3522	1.4236	1.4850	1.5388
2			0.0000	0.2970	0.4950	0.6418	0.7574	0.8522	0.9323	1.0014
3					0.0000	0.2015	0.3527	0.4728	0.5720	0.6561
4							0.0000	0.1525	0.2745	0.3758
5									0.0000	0.1227

k	11	12	13	14	15	n 16	17	18	19	20
1	1.5864	1.6292	1.6680	1.7034	1.7359	1.7660	1.7939	1.8200	1.8445	1.8675
2	1.0619	1.1157	1.1641	1.2079	1.2479	1.2847	1.3188	1.3504	1.3799	1.4076
3	0.7288	0.7928	0.8498	0.9011	0.9477	0.9903	1.0295	1.0657	1.0995	1.1309
4	0.4620	0.5368	0.6029	0.6618	0.7149	0.7632	0.8074	0.8481	0.8859	0.9210
5	0.2249	0.3122	0.3883	0.4556	0.5157	0.5700	0.6195	0.6648	0.7066	0.7454
6	0.0000	0.1026	0.1905	0.2673	0.3353	0.3962	0.4513	0.5016	0.5477	0.5903
7			0.0000	0.0882	0.1653	0.2338	0.2952	0.3508	0.4016	0.4483
8					0.0000	0.0773	0.1460	0.2077	0.2637	0.3149
9							0.0000	0.0688	0.1307	0.1870
10									0.0000	0.0620

k	21	22	23	24	25	n 26	27	28	29	30
1	1.8892	1.9097	1.9292	1.9477	1.9653	1.9822	1.9983	2.0137	2.0285	2.0428
2	1.4336	1.4582	1.4814	1.5034	1.5243	1.5442	1.5633	1.5815	1.5989	1.6156
3	1.1605	1.1882	1.2144	1.2392	1.2628	1.2851	1.3064	1.3267	1.3462	1.3648
4	0.9538	0.9846	1.0136	1.0409	1.0668	1.0914	1.1147	1.1370	1.1582	1.1786
5	0.7815	0.8153	0.8470	0.8768	0.9050	0.9317	0.9570	0.9812	1.0041	1.0261
6	0.6298	0.6667	0.7012	0.7335	0.7641	0.7929	0.8202	0.8462	0.8708	0.8944
7	0.4915	0.5316	0.5690	0.6040	0.6369	0.6679	0.6973	0.7251	0.7515	0.7767
8	0.3620	0.4056	0.4461	0.4839	0.5193	0.5527	0.5841	0.6138	0.6420	0.6689
9	0.2384	0.2858	0.3297	0.3705	0.4086	0.4444	0.4780	0.5098	0.5398	0.5683
10	0.1184	0.1700	0.2175	0.2616	0.3027	0.3410	0.3771	0.4110	0.4430	0.4733
11	0.0000	0.0564	0.1081	0.1558	0.2001	0.2413	0.2798	0.3160	0.3501	0.3824
12			0.0000	0.0518	0.0995	0.1439	0.1852	0.2239	0.2602	0.2945
13					0.0000	0.0478	0.0922	0.1336	0.1724	0.2088
14							0.0000	0.0444	0.0859	0.1247
15									0.0000	0.0415

Suppose that n random observations taken from $N(0, 1)$ are arranged in decreasing order of magnitude, and that the kth observation in the ordering is denoted by $z_{(k)}$ (so that $z_{(1)} > z_{(2)} > z_{(3)} \ldots > z_{(n)}$). The table gives the expected value $E(n, k)$ of $z_{(k)}$ for sample sizes up to 50 and $k \leqslant \frac{1}{2}(n+1)$. The relation

$$E(n, k) = -E(n, n+1-k)$$

gives the values for $k > \frac{1}{2}(n+1)$. In a random sample of 20 observations from $N(0, 1)$, for example, the expected value of the smallest observation is -1.8675 and that of the second largest is 1.4076.

Normal scores are useful in a graphical check of normality in a sample. If the underlying distribution is normal, a plot of the observed sample values against the corresponding normal scores should give an approximate straight line. An example is given on p. 34.

Sums of squares of normal scores are given on p. 34.

NORMAL SCORES (RANKITS) $E(n, k)$

k	31	32	33	34	35	36	37	38	39	40
1	2.0565	2.0697	2.0824	2.0947	2.1066	2.1181	2.1293	2.1401	2.1506	2.1608
2	1.6317	1.6471	1.6620	1.6764	1.6902	1.7036	1.7166	1.7291	1.7413	1.7531
3	1.3827	1.3998	1.4164	1.4323	1.4476	1.4624	1.4768	1.4906	1.5040	1.5170
4	1.1980	1.2167	1.2347	1.2520	1.2686	1.2847	1.3002	1.3151	1.3296	1.3437
5	1.0471	1.0672	1.0865	1.1051	1.1230	1.1402	1.1568	1.1728	1.1883	1.2033
6	0.9169	0.9384	0.9590	0.9789	0.9979	1.0162	1.0339	1.0509	1.0674	1.0833
7	0.8007	0.8236	0.8455	0.8666	0.8868	0.9063	0.9250	0.9430	0.9604	0.9772
8	0.6944	0.7187	0.7420	0.7643	0.7857	0.8063	0.8260	0.8451	0.8634	0.8811
9	0.5955	0.6213	0.6460	0.6695	0.6921	0.7138	0.7346	0.7547	0.7740	0.7926
10	0.5021	0.5294	0.5555	0.5804	0.6043	0.6271	0.6490	0.6701	0.6904	0.7099
11	0.4129	0.4418	0.4694	0.4957	0.5208	0.5449	0.5679	0.5900	0.6113	0.6318
12	0.3269	0.3575	0.3867	0.4144	0.4409	0.4662	0.4904	0.5136	0.5359	0.5574
13	0.2432	0.2757	0.3065	0.3358	0.3637	0.3903	0.4158	0.4401	0.4635	0.4859
14	0.1613	0.1957	0.2283	0.2592	0.2886	0.3166	0.3434	0.3689	0.3934	0.4169
15	0.0804	0.1169	0.1515	0.1842	0.2152	0.2446	0.2727	0.2995	0.3252	0.3498
16	0.0000	0.0389	0.0755	0.1101	0.1428	0.1739	0.2034	0.2316	0.2585	0.2842
17			0.0000	0.0366	0.0712	0.1040	0.1351	0.1647	0.1929	0.2199
18					0.0000	0.0346	0.0674	0.0985	0.1282	0.1564
19							0.0000	0.0328	0.0640	0.0936
20									0.0000	0.0312

k	41	42	43	44	45	46	47	48	49	50
1	2.1707	2.1803	2.1897	2.1988	2.2077	2.2164	2.2249	2.2331	2.2412	2.2491
2	1.7646	1.7757	1.7865	1.7971	1.8073	1.8173	1.8271	1.8366	1.8458	1.8549
3	1.5296	1.5419	1.5538	1.5653	1.5766	1.5875	1.5982	1.6086	1.6187	1.6286
4	1.3573	1.3705	1.3833	1.3957	1.4078	1.4196	1.4311	1.4422	1.4531	1.4637
5	1.2178	1.2319	1.2456	1.2588	1.2717	1.2842	1.2964	1.3083	1.3198	1.3311
6	1.0987	1.1136	1.1281	1.1421	1.1558	1.1690	1.1819	1.1944	1.2066	1.2185
7	0.9935	1.0092	1.0245	1.0392	1.0536	1.0675	1.0810	1.0942	1.1070	1.1195
8	0.8983	0.9148	0.9308	0.9463	0.9614	0.9760	0.9902	1.0040	1.0174	1.0304
9	0.8106	0.8279	0.8447	0.8610	0.8767	0.8920	0.9068	0.9212	0.9353	0.9489
10	0.7287	0.7469	0.7645	0.7815	0.7979	0.8139	0.8294	0.8444	0.8590	0.8732
11	0.6515	0.6705	0.6889	0.7067	0.7238	0.7405	0.7566	0.7723	0.7875	0.8023
12	0.5780	0.5979	0.6171	0.6356	0.6535	0.6709	0.6877	0.7040	0.7198	0.7351
13	0.5075	0.5283	0.5483	0.5676	0.5863	0.6044	0.6219	0.6388	0.6552	0.6712
14	0.4394	0.4611	0.4820	0.5022	0.5217	0.5405	0.5586	0.5763	0.5933	0.6099
15	0.3734	0.3960	0.4178	0.4389	0.4591	0.4787	0.4976	0.5159	0.5336	0.5508
16	0.3089	0.3326	0.3553	0.3772	0.3983	0.4187	0.4383	0.4573	0.4757	0.4935
17	0.2457	0.2704	0.2942	0.3170	0.3390	0.3602	0.3806	0.4003	0.4194	0.4379
18	0.1835	0.2093	0.2341	0.2579	0.2808	0.3029	0.3241	0.3446	0.3644	0.3836
19	0.1219	0.1490	0.1749	0.1997	0.2236	0.2465	0.2686	0.2899	0.3105	0.3304
20	0.0608	0.0892	0.1163	0.1422	0.1671	0.1910	0.2140	0.2361	0.2575	0.2781
21	0.0000	0.0297	0.0580	0.0851	0.1111	0.1360	0.1599	0.1830	0.2051	0.2265
22		0.0000	0.0283	0.0555	0.0814	0.1064	0.1303	0.1534	0.1756
23				0.0000	0.0271	0.0531	0.0781	0.1020	0.1251
24						0.0000	0.0260	0.0509	0.0749
25								0.0000	0.0250

Source: For the tables on pp. 32, 33 and the first table on p. 34, H. L. Harter, *Biometrika* **48** (1961), 151–65. This is reproduced in *Biometrika Tables for Statisticians*, vol. 2, Tables 9 and 13.

Example. The figure is a plot of sample values from N(0, 1) against normal scores. The sample values are taken from the top row of the table on p. 88.

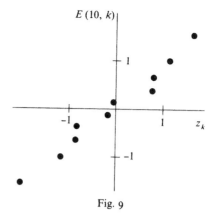

Fig. 9

SUMS OF SQUARES OF NORMAL SCORES

n	S	n	S	n	S	n	S	n	S
1	0.0000	11	8.8793	21	18.6631	31	28.5485	41	38.4727
2	0.6366	12	9.8481	22	19.6488	32	29.5397	42	39.4664
3	1.4324	13	10.8200	23	20.6354	33	30.5312	43	40.4602
4	2.2957	14	11.7945	24	21.6226	34	31.5229	44	41.4542
5	3.1950	15	12.7712	25	22.6104	35	32.5150	45	42.4485
6	4.1166	16	13.7497	26	23.5989	36	33.5073	46	43.4428
7	5.0528	17	14.7299	27	24.5880	37	34.5000	47	44.4374
8	5.9995	18	15.7114	28	25.5774	38	35.4928	48	45.4319
9	6.9539	19	16.6942	29	26.5674	39	36.4859	49	46.4268
10	7.9143	20	17.6782	30	27.5578	40	37.4792	50	47.4217

The table gives values of $S(n)$:

$$S(n) = \sum_{k=1}^{n} \{E(n, k)\}^2$$

THE STANDARDISED RANGE

n	E(W)	σ_W	n	E(W)	σ_W	n	E(W)	σ_W	n	E(W)	σ_W
			6	2.534	0.848	11	3.173	0.787	16	3.532	0.750
2	1.128	0.853	7	2.704	0.833	12	3.258	0.778	17	3.588	0.744
3	1.693	0.888	8	2.847	0.820	13	3.336	0.770	18	3.640	0.739
4	2.059	0.880	9	2.970	0.808	14	3.407	0.763	19	3.689	0.733
5	2.326	0.864	10	3.078	0.797	15	3.472	0.756	20	3.735	0.729

Source: H. L. Harter, *Annals of Mathematical Statistics* **31** (1960), 1122–47; *Biometrika Tables for Statisticians*, vol. 1 (3rd edn), Table 20.

If R is the range of a random sample of size n taken from a normal population with standard deviation σ, the *standardised range* is $W = R/\sigma$. The table gives expected values $E(W)$ and standard deviations σ_W of W. If σ is unknown, an estimate is obtained from $R/E(W)$. If several samples are available a better estimate is provided by $\bar{R}/E(W)$.

Quantiles of the W-distribution can be found in the table of quantiles of the studentised range with $\nu = \infty$ (p. 37).

THE STANDARD NEGATIVE EXPONENTIAL DISTRIBUTION

u	0	1	2	3	4	5	6	7	8	9
0.0	1	.9900	.9802	.9704	.9608	.9512	.9418	.9324	.9231	.9139
0.1	.9048	.8958	.8869	.8781	.8694	.8607	.8521	.8437	.8353	.8270
0.2	.8187	.8106	.8025	.7945	.7866	.7788	.7711	.7634	.7558	.7483
0.3	.7408	.7334	.7261	.7189	.7118	.7047	.6977	.6907	.6839	.6771
0.4	.6703	.6637	.6570	.6505	.6440	.6376	.6313	.6250	.6188	.6126
0.5	.6065	.6005	.5945	.5886	.5827	.5769	.5712	.5655	.5599	.5543
0.6	.5488	.5434	.5379	.5326	.5273	.5220	.5169	.5117	.5066	.5016
0.7	.4966	.4916	.4868	.4819	.4771	.4724	.4677	.4630	.4584	.4538
0.8	.4493	.4449	.4404	.4360	.4317	.4274	.4232	.4190	.4148	.4107
0.9	.4066	.4025	.3985	.3946	.3906	.3867	.3829	.3791	.3753	.3716
1.0	.3679	.3642	.3606	.3570	.3535	.3499	.3465	.3430	.3396	.3362
1.1	.3329	.3296	.3263	.3230	.3198	.3166	.3135	.3104	.3073	.3042
1.2	.3012	.2892	.2952	.2923	.2894	.2865	.2837	.2808	.2780	.2753
1.3	.2725	.2698	.2671	.2645	.2618	.2592	.2567	.2541	.2516	.2491
1.4	.2466	.2441	.2417	.2393	.2369	.2346	.2322	.2299	.2276	.2254
1.5	.2231	.2209	.2187	.2165	.2144	.2122	.2101	.2080	.2060	.2039
1.6	.2019	.1999	.1979	.1959	.1940	.1920	.1901	.1882	.1864	.1845
1.7	.1827	.1809	.1791	.1773	.1755	.1738	.1720	.1703	.1686	.1670
1.8	.1653	.1637	.1620	.1604	.1588	.1572	.1557	.1541	.1526	.1511
1.9	.1496	.1481	.1466	.1451	.1437	.1423	.1409	.1395	.1381	.1367
2.0	.1353	.1340	.1327	.1313	.1300	.1287	.1275	.1262	.1249	.1237
2.1	.1225	.1212	.1200	.1188	.1177	.1165	.1153	.1142	.1130	.1119
2.2	.1108	.1097	.1086	.1075	.1065	.1054	.1044	.1033	.1023	.1013
2.3	.1003	.0993	.0983	.0973	.0963	.0954	.0944	.0935	.0926	.0916
2.4	.0907	.0898	.0889	.0880	.0872	.0863	.0854	.0846	.0837	.0829
2.5	.0821	.0813	.0805	.0797	.0789	.0781	.0773	.0765	.0758	.0750
2.6	.0743	.0735	.0728	.0721	.0714	.0707	.0699	.0693	.0686	.0679
2.7	.0672	.0665	.0659	.0652	.0646	.0639	.0633	.0627	.0620	.0614
2.8	.0608	.0602	.0596	.0590	.0584	.0578	.0573	.0567	.0561	.0556
2.9	.0550	.0545	.0539	.0534	.0529	.0523	.0518	.0513	.0508	.0503
3	.0498	.0450	.0408	.0369	.0334	.0302	.0273	.0247	.0224	.0202
4	.0183	.0166	.0150	.0136	.0123	.0111	.0101	$.0^2910$	$.0^2823$	$.0^2745$
5	$.0^2674$	$.0^2610$	$.0^2552$	$.0^2499$	$.0^2452$	$.0^2409$	$.0^2370$	$.0^2335$	$.0^2303$	$.0^2274$
6	$.0^2248$	$.0^2224$	$.0^2203$	$.0^2184$	$.0^2166$	$.0^2150$	$.0^2136$	$.0^2123$	$.0^2111$	$.0^2101$
7	$.0^3912$	$.0^3825$	$.0^3747$	$.0^3676$	$.0^3611$	$.0^3553$	$.0^3500$	$.0^3453$	$.0^3410$	$.0^3371$
8	$.0^3335$	$.0^3304$	$.0^3275$	$.0^3249$	$.0^3225$	$.0^3203$	$.0^3184$	$.0^3167$	$.0^3151$	$.0^3136$
9	$.0^3123$	$.0^3112$	$.0^3101$	$.0^4914$	$.0^4827$	$.0^4749$	$.0^4677$	$.0^4613$	$.0^4555$	$.0^4502$
10	$.0^4454$	$.0^4411$	$.0^4372$	$.0^4336$	$.0^4304$	$.0^4275$	$.0^4249$	$.0^4225$	$.0^4204$	$.0^4185$
11	$.0^4167$	$.0^4151$	$.0^4137$	$.0^4124$	$.0^4112$	$.0^4101$	$.0^5917$	$.0^5829$	$.0^5750$	$.0^5679$
12	$.0^5614$	$.0^5556$	$.0^5503$	$.0^5455$	$.0^5412$	$.0^5373$	$.0^5337$	$.0^5305$	$.0^5276$	$.0^5250$
13	$.0^5226$	$.0^5205$	$.0^5185$	$.0^5167$	$.0^5152$	$.0^5137$	$.0^5124$	$.0^5112$	$.0^5102$	$.0^6919$
14	$.0^6832$	$.0^6752$	$.0^6681$	$.0^6616$	$.0^6557$	$.0^6504$	$.0^6456$	$.0^6413$	$.0^6374$	$.0^6338$

A variable U with range $(0, \infty)$ and probability density e^{-u} has the *standard negative exponential distribution*. The mean and variance of U are both 1. The table gives upper tail probabilities, also e^{-u}. Linear interpolation gives good accuracy. For $u \geqslant 15$ use the relation:

$$e^{-u} = \text{antilog}(-u \lg e) \approx \text{antilog}(-0.43429\,u)$$

If T has range $(0, \infty)$ and probability density $e^{-t/\tau}/\tau$, then T has a *negative exponential distribution*, with mean τ and variance τ^2. T/τ has the standard negative exponential distribution; hence

$$\text{Prob}(T > t) = \text{Prob}(T/\tau > t/\tau) = e^{-t/\tau}$$

which can be found from the above table. Quantiles can be found by inverse use of the table or, alternatively, from the formula:

$$t_{[P]} = -\tau \ln(1 - P)$$

UPPER QUANTILES $q_{[P]}$ OF THE STUDENTISED RANGE

ν	P	2	3	4	5	6	7	8	9	10	15	30	60	100
1	.90	8.93	13.4	16.4	18.5	20.1	21.5	22.6	23.6	24.5	27.6	32.5	36.9	39.9
	.95	18.0	27.0	32.8	37.1	40.4	43.1	45.4	47.4	49.1	55.4	65.1	74.0	80.0
	.99	90.0	135.	164.	186.	202.	216.	227.	237.	246.	277.	326.	370.	400.
	.995	180.	270.	328.	371.	404.	432.	454.	474.	491.	554.	652.	740.	800.
2	.90	4.13	5.73	6.77	7.54	8.14	8.63	9.05	9.41	9.72	10.9	12.7	14.4	15.5
	.95	6.08	8.33	9.80	10.9	11.7	12.4	13.0	13.5	14.0	15.6	18.3	20.7	22.3
	.99	14.0	19.0	22.3	24.7	26.6	28.2	29.5	30.7	31.7	35.4	41.3	46.7	50.4
	.995	19.9	27.0	31.6	35.0	37.7	39.9	41.8	43.5	44.9	50.2	58.5	66.1	71.3
3	.90	3.33	4.47	5.20	5.74	6.16	6.51	6.81	7.06	7.29	8.12	9.44	10.6	11.5
	.95	4.50	5.91	6.82	7.50	8.04	8.48	8.85	9.18	9.46	10.5	12.2	13.8	14.8
	.99	8.26	10.6	12.2	13.3	14.2	15.0	15.6	16.2	16.7	18.5	21.4	24.1	26.0
	.995	10.5	13.5	15.4	16.9	18.1	19.0	19.8	20.5	21.1	23.5	27.1	30.5	32.9
4	.90	3.01	3.98	4.59	5.03	5.39	5.68	5.93	6.14	6.33	7.02	8.13	9.16	9.86
	.95	3.93	5.04	5.76	6.29	6.71	7.05	7.35	7.60	7.83	8.66	10.0	11.2	12.1
	.99	6.51	8.12	9.17	9.96	10.6	11.1	11.5	11.9	12.3	13.5	15.6	17.5	18.8
	.995	7.92	9.81	11.1	12.0	12.7	13.3	13.9	14.3	14.7	16.2	18.7	20.9	22.5
5	.90	2.85	3.72	4.26	4.66	4.98	5.24	5.46	5.65	5.82	6.44	7.43	8.35	8.99
	.95	3.63	4.60	5.22	5.67	6.03	6.33	6.58	6.80	6.99	7.72	8.87	9.95	10.7
	.99	5.70	6.98	7.80	8.42	8.91	9.32	9.67	9.97	10.2	11.2	12.9	14.4	15.4
	.995	6.75	8.20	9.14	9.85	10.4	10.9	11.3	11.6	11.9	13.1	15.0	16.7	17.9
6	.90	2.75	3.56	4.06	4.43	4.73	4.97	5.17	5.34	5.50	6.07	7.00	7.85	8.44
	.95	3.46	4.34	4.90	5.30	5.63	5.89	6.12	6.32	6.49	7.14	8.19	9.16	9.84
	.99	5.24	6.33	7.03	7.56	7.97	8.32	8.61	8.87	9.10	9.95	11.3	12.6	13.5
	.995	6.10	7.31	8.09	8.67	9.13	9.52	9.85	10.1	10.4	11.4	12.9	14.4	15.4
7	.90	2.68	3.45	3.93	4.28	4.55	4.78	4.97	5.14	5.28	5.83	6.69	7.50	8.06
	.95	3.34	4.16	4.68	5.06	5.36	5.61	5.81	6.00	6.16	6.76	7.73	8.63	9.26
	.99	4.95	5.92	6.54	7.00	7.37	7.68	7.94	8.17	8.37	9.12	10.4	11.5	12.3
	.995	5.70	6.75	7.43	7.93	8.34	8.67	8.96	9.21	9.43	10.3	11.6	12.9	13.8
8	.90	2.63	3.37	3.83	4.17	4.43	4.65	4.83	4.99	5.13	5.64	6.47	7.24	7.78
	.95	3.26	4.04	4.53	4.89	5.17	5.40	5.60	5.77	5.92	6.48	7.39	8.25	8.84
	.99	4.75	5.63	6.20	6.62	6.96	7.24	7.47	7.68	7.86	8.55	9.68	10.7	11.5
	.995	5.42	6.37	6.98	7.43	7.80	8.10	8.35	8.58	8.78	9.53	10.8	11.9	12.7
9	.90	2.59	3.32	3.76	4.08	4.34	4.54	4.72	4.87	5.01	5.51	6.31	7.05	7.57
	.95	3.20	3.95	4.41	4.76	5.02	5.24	5.43	5.60	5.74	6.28	7.14	7.96	8.53
	.99	4.60	5.43	5.96	6.35	6.66	6.91	7.13	7.32	7.49	8.13	9.18	10.2	10.9
	.995	5.22	6.10	6.66	7.07	7.40	7.68	7.91	8.12	8.30	8.99	10.1	11.2	12.0
10	.90	2.56	3.27	3.70	4.02	4.26	4.46	4.64	4.78	4.91	5.40	6.17	6.89	7.40
	.95	3.15	3.88	4.33	4.65	4.91	5.12	5.30	5.46	5.60	6.11	6.95	7.73	8.28
	.99	4.48	5.27	5.77	6.14	6.43	6.67	6.87	7.05	7.21	7.81	8.79	9.73	10.39
	.995	5.06	5.89	6.41	6.80	7.11	7.36	7.58	7.77	7.94	8.58	9.63	10.64	11.35
12	.90	2.52	3.20	3.62	3.92	4.16	4.35	4.51	4.65	4.78	5.24	5.98	6.66	7.14
	.95	3.08	3.77	4.20	4.51	4.75	4.95	5.12	5.26	5.39	5.88	6.66	7.39	7.91
	.99	4.32	5.05	5.50	5.84	6.10	6.32	6.51	6.67	6.81	7.36	8.25	9.09	9.69
	.995	4.85	5.60	6.07	6.42	6.69	6.92	7.12	7.29	7.44	8.01	8.95	9.85	10.49
15	.90	2.48	3.14	3.54	3.83	4.05	4.23	4.39	4.52	4.64	5.08	5.78	6.43	6.89
	.95	3.01	3.67	4.08	4.37	4.59	4.78	4.94	5.08	5.20	5.65	6.38	7.06	7.55
	.99	4.17	4.84	5.25	5.56	5.80	5.99	6.16	6.31	6.44	6.93	7.73	8.49	9.03
	.995	4.65	5.32	5.75	6.06	6.31	6.51	6.68	6.84	6.97	7.48	8.31	9.11	9.68
20	.90	2.44	3.08	3.46	3.74	3.95	4.12	4.27	4.40	4.51	4.92	5.59	6.20	6.63
	.95	2.95	3.58	3.96	4.23	4.44	4.62	4.77	4.90	5.01	5.43	6.10	6.74	7.19
	.99	4.02	4.64	5.02	5.29	5.51	5.69	5.84	5.97	6.09	6.52	7.24	7.92	8.40
	.995	4.46	5.07	5.45	5.73	5.95	6.13	6.28	6.42	6.54	6.98	7.71	8.42	8.92

(right margin column, against rows 9–20)

60	ν
6	
5	
4	
3	

Sources: For the table on p. 36 and the first table on p. 37, H. L. Harter, *Annals of Mathematical Statistics* **31** (1960), 1122–47. For the second table on p. 37, *Handbook of Statistical Tables* (D. B. Owen).

ν	P	2	3	4	5	6	7	8	9	10	15	30	60	100	$\dfrac{60}{\nu}$
30	.90	2.40	3.02	3.39	3.65	3.85	4.02	4.15	4.27	4.38	4.77	5.39	5.97	6.37	2
	.95	2.89	3.49	3.84	4.10	4.30	4.46	4.60	4.72	4.82	5.21	5.83	6.42	6.83	
	.99	3.89	4.45	4.80	5.05	5.24	5.40	5.54	5.65	5.76	6.14	6.77	7.37	7.80	
	.995	4.28	4.84	5.18	5.43	5.62	5.78	5.91	6.03	6.13	6.52	7.15	7.76	8.19	
60	.90	2.36	2.96	3.31	3.56	3.75	3.91	4.04	4.15	4.25	4.62	5.20	5.73	6.10	1
	.95	2.83	3.40	3.74	3.98	4.16	4.31	4.44	4.55	4.65	5.00	5.57	6.09	6.46	
	.99	3.76	4.28	4.59	4.82	4.99	5.13	5.25	5.36	5.45	5.78	6.33	6.84	7.21	
	.995	4.12	4.62	4.93	5.15	5.32	5.45	5.57	5.67	5.76	6.09	6.63	7.14	7.50	
∞	.90	2.33	2.90	3.24	3.48	3.66	3.81	3.93	4.04	4.13	4.47	5.00	5.48	5.81	0
	.95	2.77	3.31	3.63	3.86	4.03	4.17	4.29	4.39	4.47	4.80	5.30	5.76	6.08	
	.99	3.64	4.12	4.40	4.60	4.76	4.88	4.99	5.08	5.16	5.45	5.91	6.34	6.64	
	.995	3.97	4.42	4.69	4.89	5.03	5.15	5.25	5.34	5.42	5.70	6.15	6.56	6.85	
	$(60/n)^{1/8}$										1.251	1.189	1.091	1.000	0.938	

LOWER QUANTILES $q_{[P]}$ OF THE STUDENTISED RANGE

ν	P	2	3	4	5	6	7	8	9	10	15	30	60	100	$\left(\dfrac{30}{\nu}\right)^{1/4}$
1	.01	.022	.192	.380	.531	.650	.746	.825	.891	.949	1.15	1.45	1.70	1.86	2.340
	.05	.111	.443	.698	.883	1.02	1.14	1.23	1.31	1.37	1.62	1.97	2.28	2.49	
	.10	.224	.653	.945	1.15	1.31	1.44	1.54	1.63	1.70	1.98	2.39	2.75	2.99	
3	.01	.019	.191	.408	.598	.753	.882	.989	1.08	1.16	1.44	1.85	2.20	2.42	1.778
	.05	.096	.435	.729	.953	1.13	1.27	1.39	1.49	1.57	1.88	2.34	2.73	2.99	
	.10	.193	.629	.956	1.20	1.38	1.53	1.66	1.76	1.85	2.18	2.67	3.09	3.37	
5	.01	.019	.191	.417	.620	.789	.932	1.05	1.15	1.24	1.57	2.04	2.43	2.69	1.565
	.05	.093	.434	.739	.978	1.17	1.32	1.45	1.56	1.65	1.99	2.50	2.93	3.21	
	.10	.187	.625	.964	1.22	1.41	1.57	1.71	1.82	1.92	2.27	2.80	3.26	3.56	
10	.01	.018	.191	.425	.640	.824	.981	1.11	1.23	1.33	1.70	2.25	2.71	3.01	1.316
	.05	.091	.433	.748	1.00	1.20	1.37	1.51	1.63	1.74	2.12	2.68	3.17	3.49	
	.10	.182	.622	.971	1.24	1.45	1.62	1.76	1.88	1.99	2.37	2.95	3.45	3.79	6
15	.01	.018	.191	.428	.648	.838	1.00	1.14	1.26	1.37	1.76	2.35	2.85	3.18	
	.05	.090	.432	.752	1.01	1.22	1.39	1.54	1.66	1.77	2.17	2.77	3.29	3.63	
	.10	.181	.621	.973	1.24	1.46	1.63	1.78	1.91	2.02	2.42	3.02	3.55	3.90	4
20	.01	.018	.191	.429	.652	.845	1.01	1.16	1.28	1.39	1.80	2.42	2.94	3.28	
	.05	.090	.432	.754	1.02	1.23	1.40	1.55	1.68	1.79	2.20	2.82	3.36	3.71	
	.10	.180	.620	.975	1.25	1.47	1.64	1.79	1.92	2.04	2.45	3.07	3.61	3.97	3
60	.01	.018	.191	.432	.660	.861	1.04	1.19	1.32	1.44	1.88	2.57	3.18	3.57	
	.05	.089	.432	.758	1.02	1.24	1.43	1.58	1.72	1.84	2.28	2.95	3.54	3.93	
	.10	.179	.619	.978	1.26	1.48	1.67	1.82	1.96	2.07	2.51	3.17	3.76	4.15	1
∞	.01	.018	.191	.434	.665	.870	1.05	1.21	1.34	1.47	1.93	2.68	3.35	3.81	
	.05	.089	.431	.760	1.03	1.25	1.44	1.60	1.74	1.86	2.32	3.03	3.68	4.11	
	.10	.178	.618	.979	1.26	1.49	1.68	1.84	1.97	2.09	2.54	3.24	3.86	4.28	0
	$(60/n)^{1/8}$										1.251	1.189	1.091	1.000	0.938	$\dfrac{60}{\nu}$

If R is the range of a random sample of n observations taken from a normal population, and s^2 is an independent estimate of the variance of the population based on ν degrees of freedom (as explained on p. 46), $q = R/s$ is the *studentised range*. The tables give quantiles of the sampling distribution.

Interpolation with respect to n should be linear in $(60/n)^{1/8}$, or $\sqrt{\sqrt{\sqrt{(60/n)}}}$. Interpolation with respect to ν should be linear in $(30/\nu)^{1/4}$ (or $\sqrt{\sqrt{(30/\nu)}}$) for $\nu < 10$, and linear in $60/\nu$ for $\nu > 10$.

QUANTILES $w_{[P]}$ OF CHI-SQUARE DISTRIBUTIONS $\chi^2(\nu)$

| P | .005 | .010 | .025 | .05 | .10 | .25 | .50 | .75 | .90 | .95 | .975 | .990 | .995 | .999 |
Q	.995	.990	.975	.95	.90	.75	.50	.25	.10	.05	.025	.010	.005	.001
ν														
1	$.0^4393$	$.0^3157$	$.0^3982$	$.0^2393$	.0158	0.102	0.455	1.323	2.706	3.841	5.024	6.635	7.879	10.83
2	.0100	.0201	.0506	0.103	0.211	0.575	1.386	2.773	4.605	5.991	7.378	9.210	10.60	13.82
3	.0717	0.115	0.216	0.352	0.584	1.213	2.366	4.108	6.251	7.815	9.348	11.34	12.84	16.27
4	0.207	0.297	0.484	0.711	1.064	1.923	3.357	5.385	7.779	9.488	11.14	13.28	14.86	18.47
5	0.412	0.554	0.831	1.145	1.610	2.675	4.351	6.626	9.236	11.07	12.83	15.09	16.75	20.52
6	0.676	0.872	1.237	1.635	2.204	3.455	5.348	7.841	10.64	12.59	14.45	16.81	18.55	22.46
7	0.989	1.239	1.690	2.167	2.833	4.255	6.346	9.037	12.02	14.07	16.01	18.48	20.28	24.32
8	1.344	1.646	2.180	2.733	3.490	5.071	7.344	10.22	13.36	15.51	17.53	20.09	21.95	26.12
9	1.735	2.088	2.700	3.325	4.168	5.899	8.343	11.39	14.68	16.92	19.02	21.67	23.59	27.88
10	2.156	2.558	3.247	3.940	4.865	6.737	9.342	12.55	15.99	18.31	20.48	23.21	25.19	29.59
11	2.603	3.053	3.816	4.575	5.578	7.584	10.34	13.70	17.28	19.68	21.92	24.73	26.76	31.26
12	3.074	3.571	4.404	5.226	6.304	8.438	11.34	14.85	18.55	21.03	23.34	26.22	28.30	32.91
13	3.565	4.107	5.009	5.892	7.042	9.299	12.34	15.98	19.81	22.36	24.74	27.69	29.82	34.53
14	4.075	4.660	5.629	6.571	7.790	10.17	13.34	17.12	21.06	23.68	26.12	29.14	31.32	36.12
15	4.601	5.229	6.262	7.261	8.547	11.04	14.34	18.25	22.31	25.00	27.49	30.58	32.80	37.70
16	5.142	5.812	6.908	7.962	9.312	11.91	15.34	19.37	23.54	26.30	28.85	32.00	34.27	39.25
17	5.697	6.408	7.564	8.672	10.09	12.79	16.34	20.49	24.77	27.59	30.19	33.41	35.72	40.79
18	6.265	7.015	8.231	9.390	10.86	13.68	17.34	21.60	25.99	28.87	31.53	34.81	37.16	42.31
19	6.844	7.633	8.907	10.12	11.65	14.56	18.34	22.72	27.20	30.14	32.85	36.19	38.58	43.82
20	7.434	8.260	9.591	10.85	12.44	15.45	19.34	23.83	28.41	31.41	34.17	37.57	40.00	45.31
21	8.034	8.897	10.28	11.59	13.24	16.34	20.34	24.93	29.62	32.67	35.48	38.93	41.40	46.80
22	8.643	9.542	10.98	12.34	14.04	17.24	21.34	26.04	30.81	33.92	36.78	40.29	42.80	48.27
23	9.260	10.20	11.69	13.09	14.85	18.14	22.34	27.14	32.01	35.17	38.08	41.64	44.18	49.73
24	9.886	10.86	12.40	13.85	15.66	19.04	23.34	28.24	33.20	36.42	39.36	42.98	45.56	51.18
25	10.52	11.52	13.12	14.61	16.47	19.94	24.34	29.34	34.38	37.65	40.65	44.31	46.93	52.62
26	11.16	12.20	13.84	15.38	17.29	20.84	25.34	30.43	35.56	38.89	41.92	45.64	48.29	54.05
27	11.81	12.88	14.57	16.15	18.11	21.75	26.34	31.53	36.74	40.11	43.19	46.96	49.64	55.48
28	12.46	13.56	15.31	16.93	18.94	22.66	27.34	32.62	37.92	41.34	44.46	48.28	50.99	56.89
29	13.12	14.26	16.05	17.71	19.77	23.57	28.34	33.71	39.09	42.56	45.72	49.59	52.34	58.30
30	13.79	14.95	16.79	18.49	20.60	24.48	29.34	34.80	40.26	43.77	46.98	50.89	53.67	59.70
40	20.71	22.16	24.43	26.51	29.05	33.66	39.34	45.62	51.81	55.76	59.34	63.69	66.77	73.40
50	27.99	29.71	32.36	34.76	37.69	42.94	49.33	56.33	63.17	67.50	71.42	76.15	79.49	86.66
60	35.53	37.48	40.48	43.19	46.46	52.29	59.33	66.98	74.40	79.08	83.30	88.38	91.95	99.61
70	43.28	45.44	48.76	51.74	55.33	61.70	69.33	77.58	85.53	90.53	95.02	100.4	104.2	112.3
80	51.17	53.54	57.15	60.39	64.28	71.14	79.33	88.13	96.58	101.9	106.6	112.3	116.3	124.8
90	59.20	61.75	65.65	69.13	73.29	80.62	89.33	98.65	107.6	113.1	118.1	124.1	128.3	137.2
100	67.33	70.06	74.22	77.93	82.36	90.13	99.33	109.1	118.5	124.3	129.6	135.8	140.2	149.4

Source: *Biometrika Tables for Statisticians*, vol. 1 (3rd edn), Table 8.

If $z_1, z_2, \ldots, z_\nu$ is a random sample from the distribution $N(0, 1)$, the sampling distribution of

$$w = z_1^2 + z_2^2 + \ldots + z_\nu^2$$

is $\chi^2(\nu)$, the *chi-square distribution on ν degrees of freedom*. The mean is ν; the variance is 2ν.

The tabulated function is $w_{[P]}$: $\text{Prob}(w < w_{[P]}) = P$ and $\text{Prob}(w > w_{[P]}) = Q = 1 - P$. Interpolation with respect to ν (for $\nu > 30$) gives adequate values, but may lead to errors of several units in the second decimal place.

For $\nu > 100$

$$z = \sqrt{(2w)} - \sqrt{(2\nu - 1)} \sim N(0, 1)$$

approximately. Quantiles are given by

$$w_{[P]} \approx \tfrac{1}{2}\{z_{[P]} + \sqrt{(2\nu - 1)}\}^2$$

where $z_{[P]}$ is a quantile of $N(0, 1)$.

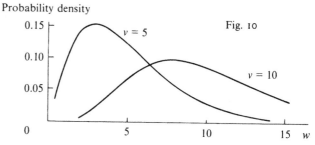

Probability density

Fig. 10

$\nu = 5$

$\nu = 10$

UPPER QUANTILES $t_{[P]}$ OF t-DISTRIBUTIONS $t(\nu)$

P	.75	.90	.95	.975	.99	.995	.9975	.999	.9995	
Q	.25	.10	.05	.025	.01	.005	.0025	.001	.0005	
$2Q$	.50	.20	.10	.050	.02	.010	.0050	.002	.0010	
$\nu=1$	1.000	3.078	6.314	12.71	31.82	63.66	127.3	318.3	636.6	
2	0.816	1.886	2.920	4.303	6.965	9.925	14.09	22.33	31.60	
3	0.765	1.638	2.353	3.182	4.541	5.841	7.453	10.21	12.92	
4	0.741	1.533	2.132	2.776	3.747	4.604	5.598	7.173	8.610	
5	0.727	1.476	2.015	2.571	3.365	4.032	4.773	5.893	6.869	
6	0.718	1.440	1.943	2.447	3.143	3.707	4.317	5.208	5.959	
7	0.711	1.415	1.895	2.365	2.998	3.499	4.029	4.785	5.408	
8	0.706	1.397	1.860	2.306	2.896	3.355	3.833	4.501	5.041	
9	0.703	1.383	1.833	2.262	2.821	3.250	3.690	4.297	4.781	
10	0.700	1.372	1.812	2.228	2.764	3.169	3.581	4.144	4.587	
11	0.697	1.363	1.796	2.201	2.718	3.106	3.497	4.025	4.437	
12	0.695	1.356	1.782	2.179	2.681	3.055	3.428	3.930	4.318	
13	0.694	1.350	1.771	2.160	2.650	3.012	3.372	3.852	4.221	
14	0.692	1.345	1.761	2.145	2.624	2.977	3.326	3.787	4.140	
15	0.691	1.341	1.753	2.131	2.602	2.947	3.286	3.733	4.073	
16	0.690	1.337	1.746	2.120	2.583	2.921	3.252	3.686	4.015	
17	0.689	1.333	1.740	2.110	2.567	2.898	3.222	3.646	3.965	
18	0.688	1.330	1.734	2.101	2.552	2.878	3.197	3.610	3.922	
19	0.688	1.328	1.729	2.093	2.539	2.861	3.174	3.579	3.883	
20	0.687	1.325	1.725	2.086	2.528	2.845	3.153	3.552	3.850	
21	0.686	1.323	1.721	2.080	2.518	2.831	3.135	3.527	3.819	
22	0.686	1.321	1.717	2.074	2.508	2.819	3.119	3.505	3.792	
23	0.685	1.319	1.714	2.069	2.500	2.807	3.104	3.485	3.767	
24	0.685	1.318	1.711	2.064	2.492	2.797	3.091	3.467	3.745	
25	0.684	1.316	1.708	2.060	2.485	2.787	3.078	3.450	3.725	
26	0.684	1.315	1.706	2.056	2.479	2.779	3.067	3.435	3.707	
27	0.684	1.314	1.703	2.052	2.473	2.771	3.057	3.421	3.690	
28	0.683	1.313	1.701	2.048	2.467	2.763	3.047	3.408	3.674	$\dfrac{120}{\nu}$
29	0.683	1.311	1.699	2.045	2.462	2.756	3.038	3.396	3.659	
30	0.683	1.310	1.697	2.042	2.457	2.750	3.030	3.385	3.646	4
40	0.681	1.303	1.684	2.021	2.423	2.704	2.971	3.307	3.551	3
60	0.679	1.296	1.671	2.000	2.390	2.660	2.915	3.232	3.460	2
120	0.677	1.289	1.658	1.980	2.358	2.617	2.860	3.160	3.373	1
∞	0.674	1.282	1.645	1.960	2.326	2.576	2.807	3.090	3.291	0

Source: *Biometrika Tables for Statisticians*, vol. 1 (3rd edn), Table 12.

If
$$t = z/\sqrt{(w/\nu)}$$

where z and w are independent variables, $z \sim N(0, 1)$ and $w \sim \chi^2(\nu)$, the distribution of t is $t(\nu)$, the *t-distribution on ν degrees of freedom*. The mean of the distribution is 0; the variance is $\nu/(\nu-2)$.

The tabulated function is $t_{[P]}$: $\mathrm{Prob}(t < t_{[P]}) = P$, $\mathrm{Prob}(t > t_{[P]}) = Q = 1-P$, and (if $P > \frac{1}{2}$) $\mathrm{Prob}(|t| > t_{[P]}) = 2Q$. Interpolation with respect to ν (for $\nu > 30$) should be linear in $120/\nu$.

Since these distributions are symmetric about the mean $t = 0$, the lower quantiles are given (for $P < \frac{1}{2}$) by:

$$t_{[P]} = -t_{[1-P]}$$

Fig. 11 shows the form of the distribution for $\nu = 2$; the shaded area represents the tail probability Q. For large ν the distributions approximate to the normal distribution $N(0, 1)$, shown by the broken line.

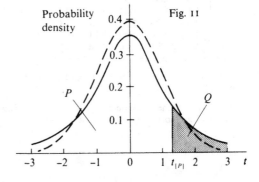

Fig. 11

UPPER QUANTILES $F_{[P]}$ OF F-DISTRIBUTIONS $F(\nu_1, \nu_2)$

ν_2	P	ν_1 1	2	3	4	5	6	7	8	9	10	12	15	20	30	60	∞
1	.9	39.9	49.5	53.6	55.8	57.2	58.2	58.9	59.4	59.9	60.2	60.7	61.2	61.7	62.3	62.8	63.3
	.95	161	200	216	225	230	234	237	239	241	242	244	246	248	250	252	254
	.975	648	800	864	900	922	937	948	957	963	969	977	985	993	1001	1010	1018
	.990	4052	5000	5403	5625	5764	5859	5928	5981	6022	6056	6106	6157	6209	6261	6313	6366
	.995	†1621	2000	2161	2250	2306	2344	2371	2393	2409	2422	2443	2463	2484	2504	2525	2546†
	.999	†4053	5000	5404	5625	5764	5859	5929	5981	6023	6056	6107	6158	6209	6261	6313	6366†
2	.9	8.53	9.00	9.16	9.24	9.29	9.33	9.35	9.37	9.38	9.39	9.41	9.42	9.44	9.46	9.47	9.49
	.95	18.5	19.0	19.2	19.2	19.3	19.3	19.4	19.4	19.4	19.4	19.4	19.4	19.4	19.5	19.5	19.5
	.975	38.5	39.0	39.2	39.2	39.3	39.3	39.4	39.4	39.4	39.4	39.4	39.4	39.4	39.5	39.5	39.5
	.990	98.5	99.0	99.2	99.2	99.3	99.3	99.4	99.4	99.4	99.4	99.4	99.4	99.4	99.5	99.5	99.5
	.995	199	199	199	199	199	199	199	199	199	199	199	199	199	199	199	199
	.999	999	999	999	999	999	999	999	999	999	999	999	999	999	999	999	999
3	.9	5.54	5.46	5.39	5.34	5.31	5.28	5.27	5.25	5.24	5.23	5.22	5.20	5.18	5.17	5.15	5.13
	.95	10.1	9.55	9.28	9.12	9.01	8.94	8.89	8.85	8.81	8.79	8.74	8.70	8.66	8.62	8.57	8.53
	.975	17.4	16.0	15.4	15.1	14.9	14.7	14.6	14.5	14.5	14.4	14.3	14.3	14.2	14.1	14.0	13.9
	.990	34.1	30.8	29.5	28.7	28.2	27.9	27.7	27.5	27.3	27.2	27.1	26.9	26.7	26.5	26.3	26.1
	.995	55.6	49.8	47.5	46.2	45.4	44.8	44.4	44.1	43.9	43.7	43.4	43.1	42.8	42.5	42.1	41.8
	.999	167	149	141	137	135	133	132	131	130	129	128	127	126	125	124	123
4	.9	4.54	4.32	4.19	4.11	4.05	4.01	3.98	3.95	3.94	3.92	3.90	3.87	3.84	3.82	3.79	3.76
	.95	7.71	6.94	6.59	6.39	6.26	6.16	6.09	6.04	6.00	5.96	5.91	5.86	5.80	5.75	5.69	5.63
	.975	12.2	10.6	9.98	9.60	9.36	9.20	9.07	8.98	8.90	8.84	8.75	8.66	8.56	8.46	8.36	8.26
	.990	21.2	18.0	16.7	16.0	15.5	15.2	15.0	14.8	14.7	14.5	14.4	14.2	14.0	13.8	13.7	13.5
	.995	31.3	26.3	24.3	23.2	22.5	22.0	21.6	21.4	21.1	21.0	20.7	20.4	20.2	19.9	19.6	19.3
	.999	74.1	61.2	56.2	53.4	51.7	50.5	49.7	49.0	48.5	48.1	47.4	46.8	46.1	45.4	44.7	44.1
5	.9	4.06	3.78	3.62	3.52	3.45	3.40	3.37	3.34	3.32	3.30	3.27	3.24	3.21	3.17	3.14	3.10
	.95	6.61	5.79	5.41	5.19	5.05	4.95	4.88	4.82	4.77	4.74	4.68	4.62	4.56	4.50	4.43	4.36
	.975	10.0	8.43	7.76	7.39	7.15	6.98	6.85	6.76	6.68	6.62	6.52	6.43	6.33	6.23	6.12	6.02
	.990	16.3	13.3	12.1	11.4	11.0	10.7	10.5	10.3	10.2	10.1	9.89	9.72	9.55	9.38	9.20	9.02
	.995	22.8	18.3	16.5	15.6	14.9	14.5	14.2	14.0	13.8	13.6	13.4	13.1	12.9	12.7	12.4	12.1
	.999	47.2	37.1	33.2	31.1	29.8	28.8	28.2	27.6	27.2	26.9	26.4	25.9	25.4	24.9	24.3	23.8
6	.9	3.78	3.46	3.29	3.18	3.11	3.05	3.01	2.98	2.96	2.94	2.90	2.87	2.84	2.80	2.76	2.72
	.95	5.99	5.14	4.76	4.53	4.39	4.28	4.21	4.15	4.10	4.06	4.00	3.94	3.87	3.81	3.74	3.67
	.975	8.81	7.26	6.60	6.23	5.99	5.82	5.70	5.60	5.52	5.46	5.37	5.27	5.17	5.07	4.96	4.85
	.990	13.7	10.9	9.78	9.15	8.75	8.47	8.26	8.10	7.98	7.87	7.72	7.56	7.40	7.23	7.06	6.88
	.995	18.6	14.5	12.9	12.0	11.5	11.1	10.8	10.6	10.4	10.3	10.0	9.81	9.59	9.36	9.12	8.88
	.999	35.5	27.0	23.7	21.9	20.8	20.0	19.5	19.0	18.7	18.4	18.0	17.6	17.1	16.7	16.2	15.7
7	.9	3.59	3.26	3.07	2.96	2.88	2.83	2.78	2.75	2.72	2.70	2.67	2.63	2.59	2.56	2.51	2.47
	.95	5.59	4.74	4.35	4.12	3.97	3.87	3.79	3.73	3.68	3.64	3.57	3.51	3.44	3.38	3.30	3.23
	.975	8.07	6.54	5.89	5.52	5.29	5.12	4.99	4.90	4.82	4.76	4.67	4.57	4.47	4.36	4.25	4.14
	.990	12.2	9.55	8.45	7.85	7.46	7.19	6.99	6.84	6.72	6.62	6.47	6.31	6.16	5.99	5.82	5.65
	.995	16.2	12.4	10.9	10.1	9.52	9.16	8.89	8.68	8.51	8.38	8.18	7.97	7.75	7.53	7.31	7.08
	.999	29.2	21.7	18.8	17.2	16.2	15.5	15.0	14.6	14.3	14.1	13.7	13.3	12.9	12.5	12.1	11.7
8	.9	3.46	3.11	2.92	2.81	2.73	2.67	2.62	2.59	2.56	2.54	2.50	2.46	2.42	2.38	2.34	2.29
	.95	5.32	4.46	4.07	3.84	3.69	3.58	3.50	3.44	3.39	3.35	3.28	3.22	3.15	3.08	3.01	2.93
	.975	7.57	6.06	5.42	5.05	4.82	4.65	4.53	4.43	4.36	4.30	4.20	4.10	4.00	3.89	3.78	3.67
	.990	11.3	8.65	7.59	7.01	6.63	6.37	6.18	6.03	5.91	5.81	5.67	5.52	5.36	5.20	5.03	4.86
	.995	14.7	11.0	9.60	8.81	8.30	7.95	7.69	7.50	7.34	7.21	7.01	6.81	6.61	6.40	6.18	5.95
	.999	25.4	18.5	15.8	14.4	13.5	12.9	12.4	12.0	11.8	11.5	11.2	10.8	10.5	10.1	9.73	9.33

† For $\nu_2 = 1$, values for $P = 0.995$ should be multiplied by 10, and values for $P = 0.999$ should be multiplied by 100.

Source: For the tables on pp. 40–3, *Biometrika Tables for Statisticians*, vol. 1 (3rd edn), Table 18. For fractional values of ν_1 and ν_2 see vol. 2, Table 4.

UPPER QUANTILES $F_{[P]}$ OF F-DISTRIBUTIONS $F(\nu_1, \nu_2)$

ν_2	P	1	2	3	4	5	6	7	8	9	10	12	15	20	30	60	∞
9	.9	3.36	3.01	2.81	2.69	2.61	2.55	2.51	2.47	2.44	2.42	2.38	2.34	2.30	2.25	2.21	2.16
	.95	5.12	4.26	3.86	3.63	3.48	3.37	3.29	3.23	3.18	3.14	3.07	3.01	2.94	2.86	2.79	2.71
	.975	7.21	5.71	5.08	4.72	4.48	4.32	4.20	4.10	4.03	3.96	3.87	3.77	3.67	3.56	3.45	3.33
	.990	10.6	8.02	6.99	6.42	6.06	5.80	5.61	5.47	5.35	5.26	5.11	4.96	4.81	4.65	4.48	4.31
	.995	13.6	10.1	8.72	7.96	7.47	7.13	6.88	6.69	6.54	6.42	6.23	6.03	5.83	5.62	5.41	5.19
	.999	22.9	16.4	13.9	12.6	11.7	11.1	10.7	10.4	10.1	9.89	9.57	9.24	8.90	8.55	8.19	7.81
10	.9	3.29	2.92	2.73	2.61	2.52	2.46	2.41	2.38	2.35	2.32	2.28	2.24	2.20	2.16	2.11	2.06
	.95	4.96	4.10	3.71	3.48	3.33	3.22	3.14	3.07	3.02	2.98	2.91	2.85	2.77	2.70	2.62	2.54
	.975	6.94	5.46	4.83	4.47	4.24	4.07	3.95	3.85	3.78	3.72	3.62	3.52	3.42	3.31	3.20	3.08
	.990	10.0	7.56	6.55	5.99	5.64	5.39	5.20	5.06	4.94	4.85	4.71	4.56	4.41	4.25	4.08	3.91
	.995	12.8	9.43	8.08	7.34	6.87	6.54	6.30	6.12	5.97	5.85	5.66	5.47	5.27	5.07	4.86	4.64
	.999	21.0	14.9	12.6	11.3	10.5	9.92	9.52	9.20	8.96	8.75	8.45	8.13	7.80	7.47	7.12	6.76
11	.9	3.23	2.86	2.66	2.54	2.45	2.39	2.34	2.30	2.27	2.25	2.21	2.17	2.12	2.08	2.03	1.97
	.95	4.84	3.98	3.59	3.36	3.20	3.09	3.01	2.95	2.90	2.85	2.79	2.72	2.65	2.57	2.49	2.40
	.975	6.72	5.26	4.63	4.28	4.04	3.88	3.76	3.66	3.59	3.53	3.43	3.33	3.23	3.12	3.00	2.88
	.990	9.65	7.21	6.22	5.67	5.32	5.07	4.89	4.74	4.63	4.54	4.40	4.25	4.10	3.94	3.78	3.60
	.995	12.2	8.91	7.60	6.88	6.42	6.10	5.86	5.68	5.54	5.42	5.24	5.05	4.86	4.65	4.44	4.23
	.999	19.7	13.8	11.6	10.3	9.58	9.05	8.66	8.35	8.12	7.92	7.63	7.32	7.01	6.68	6.35	6.00
12	.9	3.18	2.81	2.61	2.48	2.39	2.33	2.28	2.24	2.21	2.19	2.15	2.10	2.06	2.01	1.96	1.90
	.95	4.75	3.89	3.49	3.26	3.11	3.00	2.91	2.85	2.80	2.75	2.69	2.62	2.54	2.47	2.38	2.30
	.975	6.55	5.10	4.47	4.12	3.89	3.73	3.61	3.51	3.44	3.37	3.28	3.18	3.07	2.96	2.85	2.72
	.990	9.33	6.93	5.95	5.41	5.06	4.82	4.64	4.50	4.39	4.30	4.16	4.01	3.86	3.70	3.54	3.36
	.995	11.8	8.51	7.23	6.52	6.07	5.76	5.52	5.35	5.20	5.09	4.91	4.72	4.53	4.33	4.12	3.90
	.999	18.6	13.0	10.8	9.63	8.89	8.38	8.00	7.71	7.48	7.29	7.00	6.71	6.40	6.09	5.76	5.42
13	.9	3.14	2.76	2.56	2.43	2.35	2.28	2.23	2.20	2.16	2.14	2.10	2.05	2.01	1.96	1.90	1.85
	.95	4.67	3.81	3.41	3.18	3.03	2.92	2.83	2.77	2.71	2.67	2.60	2.53	2.46	2.38	2.30	2.21
	.975	6.41	4.97	4.35	4.00	3.77	3.60	3.48	3.39	3.31	3.25	3.15	3.05	2.95	2.84	2.72	2.60
	.990	9.07	6.70	5.74	5.21	4.86	4.62	4.44	4.30	4.19	4.10	3.96	3.82	3.66	3.51	3.34	3.17
	.995	11.4	8.19	6.93	6.23	5.79	5.48	5.25	5.08	4.94	4.82	4.64	4.46	4.27	4.07	3.87	3.65
	.999	17.8	12.3	10.2	9.07	8.35	7.86	7.49	7.21	6.98	6.80	6.52	6.23	5.93	5.63	5.30	4.97
14	.9	3.10	2.73	2.52	2.39	2.31	2.24	2.19	2.15	2.12	2.10	2.05	2.01	1.96	1.91	1.86	1.80
	.95	4.60	3.74	3.34	3.11	2.96	2.85	2.76	2.70	2.65	2.60	2.53	2.46	2.39	2.31	2.22	2.13
	.975	6.30	4.86	4.24	3.89	3.66	3.50	3.38	3.29	3.21	3.15	3.05	2.95	2.84	2.73	2.61	2.49
	.990	8.86	6.51	5.56	5.04	4.69	4.46	4.28	4.14	4.03	3.94	3.80	3.66	3.51	3.35	3.18	3.00
	.995	11.1	7.92	6.68	6.00	5.56	5.26	5.03	4.86	4.72	4.60	4.43	4.25	4.06	3.86	3.66	3.44
	.999	17.1	11.8	9.73	8.62	7.92	7.43	7.08	6.80	6.58	6.40	6.13	5.85	5.56	5.25	4.94	4.60
15	.9	3.07	2.70	2.49	2.36	2.27	2.21	2.16	2.12	2.09	2.06	2.02	1.97	1.92	1.87	1.82	1.76
	.95	4.54	3.68	3.29	3.06	2.90	2.79	2.71	2.64	2.59	2.54	2.48	2.40	2.33	2.25	2.16	2.07
	.975	6.20	4.77	4.15	3.80	3.58	3.41	3.29	3.20	3.12	3.06	2.96	2.86	2.76	2.64	2.52	2.40
	.990	8.68	6.36	5.42	4.89	4.56	4.32	4.14	4.00	3.89	3.80	3.67	3.52	3.37	3.21	3.05	2.87
	.995	10.8	7.70	6.48	5.80	5.37	5.07	4.85	4.67	4.54	4.42	4.25	4.07	3.88	3.69	3.48	3.26
	.999	16.6	11.3	9.34	8.25	7.57	7.09	6.74	6.47	6.26	6.08	5.81	5.54	5.25	4.95	4.64	4.31
16	.9	3.05	2.67	2.46	2.33	2.24	2.18	2.13	2.09	2.06	2.03	1.99	1.94	1.89	1.84	1.78	1.72
	.95	4.49	3.63	3.24	3.01	2.85	2.74	2.66	2.59	2.54	2.49	2.42	2.35	2.28	2.19	2.11	2.01
	.975	6.12	4.69	4.08	3.73	3.50	3.34	3.22	3.12	3.05	2.99	2.89	2.79	2.68	2.57	2.45	2.32
	.990	8.53	6.23	5.29	4.77	4.44	4.20	4.03	3.89	3.78	3.69	3.55	3.41	3.26	3.10	2.93	2.75
	.995	10.6	7.51	6.30	5.64	5.21	4.91	4.69	4.52	4.38	4.27	4.10	3.92	3.73	3.54	3.33	3.11
	.999	16.1	11.0	9.01	7.94	7.27	6.81	6.46	6.19	5.98	5.81	5.55	5.27	4.99	4.70	4.39	4.06
17	.9	3.03	2.64	2.44	2.31	2.22	2.15	2.10	2.06	2.03	2.00	1.96	1.91	1.86	1.81	1.75	1.69
	.95	4.45	3.59	3.20	2.96	2.81	2.70	2.61	2.55	2.49	2.45	2.38	2.31	2.23	2.15	2.06	1.96
	.975	6.04	4.62	4.01	3.66	3.44	3.28	3.16	3.06	2.98	2.92	2.82	2.72	2.62	2.50	2.38	2.25
	.990	8.40	6.11	5.18	4.67	4.34	4.10	3.93	3.79	3.68	3.59	3.46	3.31	3.16	3.00	2.83	2.65
	.995	10.4	7.35	6.16	5.50	5.07	4.78	4.56	4.39	4.25	4.14	3.97	3.79	3.61	3.41	3.21	2.98
	.999	15.7	10.7	8.73	7.68	7.02	6.56	6.22	5.96	5.75	5.58	5.32	5.05	4.78	4.48	4.18	3.85

UPPER QUANTILES $F_{[P]}$ OF F-DISTRIBUTIONS $F(\nu_1, \nu_2)$

ν_2	P	ν_1 1	2	3	4	5	6	7	8	9	10	12	15	20	30	60	∞
18	.9	3.01	2.62	2.42	2.29	2.20	2.13	2.08	2.04	2.00	1.98	1.93	1.89	1.84	1.78	1.72	1.66
	.95	4.41	3.55	3.16	2.93	2.77	2.66	2.58	2.51	2.46	2.41	2.34	2.27	2.19	2.11	2.02	1.92
	.975	5.98	4.56	3.95	3.61	3.38	3.22	3.10	3.01	2.93	2.87	2.77	2.67	2.56	2.44	2.32	2.19
	.990	8.29	6.01	5.09	4.58	4.25	4.01	3.84	3.71	3.60	3.51	3.37	3.23	3.08	2.92	2.75	2.57
	.995	10.2	7.21	6.03	5.37	4.96	4.66	4.44	4.28	4.14	4.03	3.86	3.68	3.50	3.30	3.10	2.87
	.999	15.4	10.4	8.49	7.46	6.81	6.35	6.02	5.76	5.56	5.39	5.13	4.87	4.59	4.30	4.00	3.67
19	.9	2.99	2.61	2.40	2.27	2.18	2.11	2.06	2.02	1.98	1.96	1.91	1.86	1.81	1.76	1.70	1.63
	.95	4.38	3.52	3.13	2.90	2.74	2.63	2.54	2.48	2.42	2.38	2.31	2.23	2.16	2.07	1.98	1.88
	.975	5.92	4.51	3.90	3.56	3.33	3.17	3.05	2.96	2.88	2.82	2.72	2.62	2.51	2.39	2.27	2.13
	.990	8.18	5.93	5.01	4.50	4.17	3.94	3.77	3.63	3.52	3.43	3.30	3.15	3.00	2.84	2.67	2.49
	.995	10.1	7.09	5.92	5.27	4.85	4.56	4.34	4.18	4.04	3.93	3.76	3.59	3.40	3.21	3.00	2.78
	.999	15.1	10.2	8.28	7.26	6.62	6.18	5.85	5.59	5.39	5.22	4.97	4.70	4.43	4.14	3.84	3.51
20	.9	2.97	2.59	2.38	2.25	2.16	2.09	2.04	2.00	1.96	1.94	1.89	1.84	1.79	1.74	1.68	1.61
	.95	4.35	3.49	3.10	2.87	2.71	2.60	2.51	2.45	2.39	2.35	2.28	2.20	2.12	2.04	1.95	1.84
	.975	5.87	4.46	3.86	3.51	3.29	3.13	3.01	2.91	2.84	2.77	2.68	2.57	2.46	2.35	2.22	2.09
	.990	8.10	5.85	4.94	4.43	4.10	3.87	3.70	3.56	3.46	3.37	3.23	3.09	2.94	2.78	2.61	2.42
	.995	9.94	6.99	5.82	5.17	4.76	4.47	4.26	4.09	3.96	3.85	3.68	3.50	3.32	3.12	2.92	2.69
	.999	14.8	9.95	8.10	7.10	6.46	6.02	5.69	5.44	5.24	5.08	4.82	4.56	4.29	4.00	3.70	3.38
21	.9	2.96	2.57	2.36	2.23	2.14	2.08	2.02	1.98	1.95	1.92	1.87	1.83	1.78	1.72	1.66	1.59
	.95	4.32	3.47	3.07	2.84	2.68	2.57	2.49	2.42	2.37	2.32	2.25	2.18	2.10	2.01	1.92	1.81
	.975	5.83	4.42	3.82	3.48	3.25	3.09	2.97	2.87	2.80	2.73	2.64	2.53	2.42	2.31	2.18	2.04
	.990	8.02	5.78	4.87	4.37	4.04	3.81	3.64	3.51	3.40	3.31	3.17	3.03	2.88	2.72	2.55	2.36
	.995	9.83	6.89	5.73	5.09	4.68	4.39	4.18	4.01	3.88	3.77	3.60	3.43	3.24	3.05	2.84	2.61
	.999	14.6	9.77	7.94	6.95	6.32	5.88	5.56	5.31	5.11	4.95	4.70	4.44	4.17	3.88	3.58	3.26
22	.9	2.95	2.56	2.35	2.22	2.13	2.06	2.01	1.97	1.93	1.90	1.86	1.81	1.76	1.70	1.64	1.57
	.95	4.30	3.44	3.05	2.82	2.66	2.55	2.46	2.40	2.34	2.30	2.23	2.15	2.07	1.98	1.89	1.78
	.975	5.79	4.38	3.78	3.44	3.22	3.05	2.93	2.84	2.76	2.70	2.60	2.50	2.39	2.27	2.14	2.00
	.990	7.95	5.72	4.82	4.31	3.99	3.76	3.59	3.45	3.35	3.26	3.12	2.98	2.83	2.67	2.50	2.31
	.995	9.73	6.81	5.65	5.02	4.61	4.32	4.11	3.94	3.81	3.70	3.54	3.36	3.18	2.98	2.77	2.55
	.999	14.4	9.61	7.80	6.81	6.19	5.76	5.44	5.19	4.99	4.83	4.58	4.33	4.06	3.78	3.48	3.15
23	.9	2.94	2.55	2.34	2.21	2.11	2.05	1.99	1.95	1.92	1.89	1.84	1.80	1.74	1.69	1.62	1.55
	.95	4.28	3.42	3.03	2.80	2.64	2.53	2.44	2.37	2.32	2.27	2.20	2.13	2.05	1.96	1.86	1.76
	.975	5.75	4.35	3.75	3.41	3.18	3.02	2.90	2.81	2.73	2.67	2.57	2.47	2.36	2.24	2.11	1.97
	.990	7.88	5.66	4.76	4.26	3.94	3.71	3.54	3.41	3.30	3.21	3.07	2.93	2.78	2.62	2.45	2.26
	.995	9.63	6.73	5.58	4.95	4.54	4.26	4.05	3.88	3.75	3.64	3.47	3.30	3.12	2.92	2.71	2.48
	.999	14.2	9.47	7.67	6.69	6.08	5.65	5.33	5.09	4.89	4.73	4.48	4.23	3.96	3.68	3.38	3.05
24	.9	2.93	2.54	2.33	2.19	2.10	2.04	1.98	1.94	1.91	1.88	1.83	1.78	1.73	1.67	1.61	1.53
	.95	4.26	3.40	3.01	2.78	2.62	2.51	2.42	2.36	2.30	2.25	2.18	2.11	2.03	1.94	1.84	1.73
	.975	5.72	4.32	3.72	3.38	3.15	2.99	2.87	2.78	2.70	2.64	2.54	2.44	2.33	2.21	2.08	1.94
	.990	7.82	5.61	4.72	4.22	3.90	3.67	3.50	3.36	3.26	3.17	3.03	2.89	2.74	2.58	2.40	2.21
	.995	9.55	6.66	5.52	4.89	4.49	4.20	3.99	3.83	3.69	3.59	3.42	3.25	3.06	2.87	2.66	2.43
	.999	14.0	9.34	7.55	6.59	5.98	5.55	5.23	4.99	4.80	4.64	4.39	4.14	3.87	3.59	3.29	2.97
30	.9	2.88	2.49	2.28	2.14	2.05	1.98	1.93	1.88	1.85	1.82	1.77	1.72	1.67	1.61	1.54	1.46
	.95	4.17	3.32	2.92	2.69	2.53	2.42	2.33	2.27	2.21	2.16	2.09	2.01	1.93	1.84	1.74	1.62
	.975	5.57	4.18	3.59	3.25	3.03	2.87	2.75	2.65	2.57	2.51	2.41	2.31	2.20	2.07	1.94	1.79
	.990	7.56	5.39	4.51	4.02	3.70	3.47	3.30	3.17	3.07	2.98	2.84	2.70	2.55	2.39	2.21	2.01
	.995	9.18	6.35	5.24	4.62	4.23	3.95	3.74	3.58	3.45	3.34	3.18	3.01	2.82	2.63	2.42	2.18
	.999	13.3	8.77	7.05	6.12	5.53	5.12	4.82	4.58	4.39	4.24	4.00	3.75	3.49	3.22	2.92	2.59
40	.9	2.84	2.44	2.23	2.09	2.00	1.93	1.87	1.83	1.79	1.76	1.71	1.66	1.61	1.54	1.47	1.38
	.95	4.08	3.23	2.84	2.61	2.45	2.34	2.25	2.18	2.12	2.08	2.00	1.92	1.84	1.74	1.64	1.51
	.975	5.42	4.05	3.46	3.13	2.90	2.74	2.62	2.53	2.45	2.39	2.29	2.18	2.07	1.94	1.80	1.64
	.990	7.31	5.18	4.31	3.83	3.51	3.29	3.12	2.99	2.89	2.80	2.66	2.52	2.37	2.20	2.02	1.80
	.995	8.83	6.07	4.98	4.37	3.99	3.71	3.51	3.35	3.22	3.12	2.95	2.78	2.60	2.40	2.18	1.93
	.999	12.6	8.25	6.60	5.70	5.13	4.73	4.44	4.21	4.02	3.87	3.64	3.40	3.15	2.87	2.57	2.23

UPPER QUANTILES $F_{[P]}$ OF F-DISTRIBUTIONS $F(\nu_1, \nu_2)$

ν_2	P	1	2	3	4	5	6	7	8	9	10	12	15	20	30	60	∞
60	.9	2.79	2.39	2.18	2.04	1.95	1.87	1.82	1.77	1.74	1.71	1.66	1.60	1.54	1.48	1.40	1.29
	.95	4.00	3.15	2.76	2.53	2.37	2.25	2.17	2.10	2.04	1.99	1.92	1.84	1.75	1.65	1.53	1.39
	.975	5.29	3.93	3.34	3.01	2.79	2.63	2.51	2.41	2.33	2.27	2.17	2.06	1.94	1.82	1.67	1.48
	.990	7.08	4.98	4.13	3.65	3.34	3.12	2.95	2.82	2.72	2.63	2.50	2.35	2.20	2.03	1.84	1.60
	.995	8.49	5.79	4.73	4.14	3.76	3.49	3.29	3.13	3.01	2.90	2.74	2.57	2.39	2.19	1.96	1.69
	.999	12.0	7.76	6.17	5.31	4.76	4.37	4.09	3.87	3.69	3.54	3.31	3.08	2.83	2.55	2.25	1.89
120	.9	2.75	2.35	2.13	1.99	1.90	1.82	1.77	1.72	1.68	1.65	1.60	1.55	1.48	1.41	1.32	1.19
	.95	3.92	3.07	2.68	2.45	2.29	2.17	2.09	2.02	1.96	1.91	1.83	1.75	1.66	1.55	1.43	1.25
	.975	5.15	3.80	3.23	2.89	2.67	2.52	2.39	2.30	2.22	2.16	2.05	1.94	1.82	1.69	1.53	1.31
	.990	6.85	4.79	3.95	3.48	3.17	2.96	2.79	2.66	2.56	2.47	2.34	2.19	2.03	1.86	1.66	1.38
	.995	8.18	5.54	4.50	3.92	3.55	3.28	3.09	2.93	2.81	2.71	2.54	2.37	2.19	1.98	1.75	1.43
	.999	11.4	7.32	5.79	4.95	4.42	4.04	3.77	3.55	3.38	3.24	3.02	2.78	2.53	2.26	1.95	1.54
∞	.9	2.71	2.30	2.08	1.94	1.85	1.77	1.72	1.67	1.63	1.60	1.55	1.49	1.42	1.34	1.24	
	.95	3.84	3.00	2.60	2.37	2.21	2.10	2.01	1.94	1.88	1.83	1.75	1.67	1.57	1.46	1.32	
	.975	5.02	3.69	3.12	2.79	2.57	2.41	2.29	2.19	2.11	2.05	1.94	1.83	1.71	1.57	1.39	
	.990	6.63	4.61	3.78	3.32	3.02	2.80	2.64	2.51	2.41	2.32	2.18	2.04	1.88	1.70	1.47	
	.995	7.88	5.30	4.28	3.72	3.35	3.09	2.90	2.74	2.62	2.52	2.36	2.19	2.00	1.79	1.53	
	.999	10.8	6.91	5.42	4.62	4.10	3.74	3.47	3.27	3.10	2.96	2.74	2.51	2.27	1.99	1.66	

If w_1 and w_2 are independent random variables with distributions $\chi^2(\nu_1)$ and $\chi^2(\nu_2)$ respectively, the distribution of

$$\left(\frac{w_1}{\nu_1}\right) \Big/ \left(\frac{w_2}{\nu_2}\right)$$

(the ratio of two mean squares estimating a common variance) is $F(\nu_1, \nu_2)$.

The table gives quantiles of $F(\nu_1, \nu_2)$ for $P = 0.9, 0.95, 0.975, 0.990, 0.995$ and 0.999. If $x \sim F(\nu_1, \nu_2)$, $\text{Prob}(x < F_{[P]}) = P$ and $\text{Prob}(x > F_{[P]}) = 1 - P$. For example, for $\nu_1 = 2$ and $\nu_2 = 60$ then $\text{Prob}(x > 3.15) = 1 - 0.95 = 0.05$ or 5%. Interpolation with respect to ν_1 should be linear in $60/\nu_1$; interpolation with respect to ν_2 should be linear in $120/\nu_2$.

Lower quantiles (for $P < \frac{1}{2}$) are given by:

$$F_{[P]}(\nu_1, \nu_2) = 1/F_{[1-P]}(\nu_2, \nu_1)$$

For two independent samples of sizes n_1 and n_2 drawn from normal populations with variances σ_1^2 and σ_2^2 respectively,

$$\left(\frac{\hat{\sigma}_1^2}{\sigma_1^2}\right) \Big/ \left(\frac{\hat{\sigma}_2^2}{\sigma_2^2}\right)$$

has the distribution $F(n_1 - 1, n_2 - 1)$. (For the definition and distribution of $\hat{\sigma}^2$ see p. 46.) If $\sigma_1^2 = \sigma_2^2$ the ratio becomes $\hat{\sigma}_1^2/\hat{\sigma}_2^2$.

FACTORS D FOR DUNCAN'S MULTIPLE RANGE TEST

ν	α	2	3	4	5	6	7	8	9	10	12	15	20	30	60	100	$60/\nu$
1	.10	8.93	8.93	8.93	8.93	8.93	8.93	8.93	8.93	8.93	8.93	8.93	8.93	8.93	8.93	8.93	
	.05	18.0	18.0	18.0	18.0	18.0	18.0	18.0	18.0	18.0	18.0	18.0	18.0	18.0	18.0	18.0	
	.01	90.0	90.0	90.0	90.0	90.0	90.0	90.0	90.0	90.0	90.0	90.0	90.0	90.0	90.0	90.0	
	.005	180.	180.	180.	180.	180.	180.	180.	180.	180.	180.	180.	180.	180.	180.	180.	
2	.10	4.13	4.13	4.13	4.13	4.13	4.13	4.13	4.13	4.13	4.13	4.13	4.13	4.13	4.13	4.13	
	.05	6.08	6.08	6.08	6.08	6.08	6.08	6.08	6.08	6.08	6.08	6.08	6.08	6.08	6.08	6.08	
	.01	14.0	14.0	14.0	14.0	14.0	14.0	14.0	14.0	14.0	14.0	14.0	14.0	14.0	14.0	14.0	
	.005	19.9	19.9	19.9	19.9	19.9	19.9	19.9	19.9	19.9	19.9	19.9	19.9	19.9	19.9	19.9	
3	.10	3.33	3.33	3.33	3.33	3.33	3.33	3.33	3.33	3.33	3.33	3.33	3.33	3.33	3.33	3.33	
	.05	4.50	4.52	4.52	4.52	4.52	4.52	4.52	4.52	4.52	4.52	4.52	4.52	4.52	4.52	4.52	
	.01	8.26	8.32	8.32	8.32	8.32	8.32	8.32	8.32	8.32	8.32	8.32	8.32	8.32	8.32	8.32	
	.005	10.5	10.6	10.6	10.6	10.6	10.6	10.6	10.6	10.6	10.6	10.6	10.6	10.6	10.6	10.6	
4	.10	3.01	3.07	3.08	3.08	3.08	3.08	3.08	3.08	3.08	3.08	3.08	3.08	3.08	3.08	3.08	
	.05	3.93	4.01	4.03	4.03	4.03	4.03	4.03	4.03	4.03	4.03	4.03	4.03	4.03	4.03	4.03	
	.01	6.51	6.68	6.74	6.76	6.76	6.76	6.76	6.76	6.76	6.76	6.76	6.76	6.76	6.76	6.76	
	.005	7.92	8.13	8.21	8.24	8.24	8.24	8.24	8.24	8.24	8.24	8.24	8.24	8.24	8.24	8.24	
5	.10	2.85	2.93	2.96	2.97	2.97	2.97	2.97	2.97	2.97	2.97	2.97	2.97	2.97	2.97	2.97	
	.05	3.63	3.75	3.80	3.81	3.81	3.81	3.81	3.81	3.81	3.81	3.81	3.81	3.81	3.81	3.81	
	.01	5.70	5.89	5.99	6.04	6.06	6.07	6.07	6.07	6.07	6.07	6.07	6.07	6.07	6.07	6.07	
	.005	6.75	6.98	7.10	7.17	7.20	7.22	7.23	7.23	7.23	7.23	7.23	7.23	7.23	7.23	7.23	
6	.10	2.75	2.85	2.89	2.91	2.91	2.91	2.91	2.91	2.91	2.91	2.91	2.91	2.91	2.91	2.91	
	.05	3.46	3.59	3.65	3.68	3.69	3.70	3.70	3.70	3.70	3.70	3.70	3.70	3.70	3.70	3.70	
	.01	5.24	5.44	5.55	5.61	5.65	5.68	5.69	5.70	5.70	5.70	5.70	5.70	5.70	5.70	5.70	
	.005	6.10	6.33	6.47	6.55	6.60	6.63	6.66	6.67	6.68	6.68	6.68	6.68	6.68	6.68	6.68	
7	.10	2.68	2.78	2.84	2.86	2.88	2.88	2.88	2.88	2.88	2.88	2.88	2.88	2.88	2.88	2.88	
	.05	3.34	3.48	3.55	3.59	3.61	3.62	3.63	3.63	3.63	3.63	3.63	3.63	3.63	3.63	3.63	
	.01	4.95	5.14	5.26	5.33	5.38	5.42	5.44	5.45	5.46	5.47	5.47	5.47	5.47	5.47	5.47	
	.005	5.70	5.92	6.06	6.14	6.21	6.25	6.28	6.30	6.32	6.34	6.34	6.34	6.34	6.34	6.34	
8	.10	2.63	2.74	2.80	2.83	2.85	2.86	2.86	2.86	2.86	2.86	2.86	2.86	2.86	2.86	2.86	
	.05	3.26	3.40	3.47	3.52	3.55	3.57	3.57	3.58	3.58	3.58	3.58	3.58	3.58	3.58	3.58	
	.01	4.75	4.94	5.06	5.13	5.19	5.23	5.26	5.28	5.29	5.31	5.32	5.32	5.32	5.32	5.32	
	.005	5.42	5.64	5.77	5.86	5.93	5.98	6.01	6.04	6.06	6.09	6.11	6.12	6.12	6.12	6.12	
9	.10	2.59	2.71	2.77	2.81	2.83	2.84	2.84	2.85	2.85	2.85	2.85	2.85	2.85	2.85	2.85	$60/\nu$
	.05	3.20	3.34	3.42	3.47	3.50	3.52	3.54	3.54	3.55	3.55	3.55	3.55	3.55	3.55	3.55	
	.01	4.60	4.79	4.91	4.99	5.04	5.09	5.12	5.14	5.16	5.18	5.20	5.21	5.21	5.21	5.21	
	.005	5.22	5.43	5.56	5.66	5.72	5.78	5.81	5.85	5.87	5.91	5.94	5.96	5.96	5.96	5.96	
10	.10	2.56	2.68	2.75	2.79	2.81	2.83	2.83	2.84	2.84	2.84	2.84	2.84	2.84	2.84	2.84	6
	.05	3.15	3.29	3.38	3.43	3.46	3.49	3.50	3.52	3.52	3.53	3.53	3.53	3.53	3.53	3.53	
	.01	4.48	4.67	4.79	4.87	4.93	4.97	5.01	5.04	5.06	5.09	5.11	5.12	5.12	5.12	5.12	
	.005	5.06	5.27	5.40	5.50	5.57	5.62	5.66	5.69	5.72	5.76	5.80	5.83	5.84	5.84	5.84	
12	.10	2.52	2.64	2.71	2.76	2.79	2.81	2.82	2.83	2.83	2.83	2.83	2.83	2.83	2.83	2.83	5
	.05	3.08	3.22	3.31	3.37	3.41	3.44	3.46	3.47	3.48	3.50	3.50	3.50	3.50	3.50	3.50	
	.01	4.32	4.50	4.62	4.71	4.77	4.81	4.85	4.88	4.91	4.94	4.98	5.01	5.01	5.01	5.01	
	.005	4.85	5.05	5.18	5.27	5.34	5.40	5.44	5.47	5.50	5.55	5.60	5.64	5.67	5.67	5.67	
15	.10	2.48	2.60	2.68	2.73	2.76	2.79	2.80	2.82	2.82	2.83	2.83	2.83	2.83	2.83	2.83	4
	.05	3.01	3.16	3.25	3.31	3.36	3.39	3.41	3.43	3.45	3.46	3.48	3.48	3.48	3.48	3.48	
	.01	4.17	4.35	4.46	4.55	4.61	4.66	4.70	4.73	4.76	4.80	4.85	4.89	4.91	4.91	4.91	
	.005	4.65	4.84	4.96	5.05	5.12	5.18	5.23	5.26	5.30	5.35	5.40	5.46	5.51	5.52	5.52	

Sources: For the tables on pp. 44, 45, D. B. Duncan, *Biometrics* **11** (1955), 1–42; H. L. Harter, *Biometrics* **16** (1960), 671–85.

Interpolation with respect to ν, for $\nu > 10$, should be linear in $60/\nu$. Interpolation with respect to p, for $p > 10$, should be linear in $60/p$.

FACTORS D FOR DUNCAN'S MULTIPLE RANGE TEST

ν	α	2	3	4	5	6	7	8	9	10	12	15	20	30	60	100	$\dfrac{60}{\nu}$
20	.10	2.44	2.57	2.65	2.70	2.74	2.77	2.79	2.81	2.82	2.83	2.84	2.84	2.84	2.84	2.84	3
	.05	2.95	3.10	3.19	3.25	3.30	3.34	3.37	3.39	3.41	3.44	3.46	3.47	3.47	3.47	3.47	
	.01	4.02	4.20	4.31	4.39	4.46	4.51	4.55	4.59	4.62	4.66	4.72	4.77	4.82	4.83	4.83	
	.005	4.46	4.64	4.76	4.85	4.92	4.98	5.02	5.06	5.09	5.15	5.21	5.28	5.35	5.40	5.40	
30	.10	2.40	2.53	2.61	2.67	2.72	2.75	2.78	2.80	2.81	2.84	2.86	2.87	2.87	2.87	2.87	2
	.05	2.89	3.03	3.13	3.20	3.25	3.29	3.32	3.35	3.37	3.40	3.44	3.47	3.49	3.49	3.49	
	.01	3.89	4.06	4.17	4.25	4.31	4.37	4.41	4.44	4.48	4.53	4.59	4.65	4.72	4.78	4.78	
	.005	4.28	4.46	4.57	4.66	4.73	4.78	4.83	4.87	4.90	4.96	5.02	5.10	5.19	5.28	5.30	
60	.10	2.36	2.50	2.58	2.65	2.69	2.73	2.76	2.79	2.81	2.84	2.87	2.91	2.93	2.94	2.94	1
	.05	2.83	2.98	3.07	3.14	3.20	3.24	3.28	3.31	3.33	3.37	3.42	3.47	3.51	3.54	3.54	
	.01	3.76	3.92	4.03	4.11	4.17	4.23	4.27	4.31	4.34	4.39	4.46	4.53	4.62	4.73	4.76	
	.005	4.12	4.28	4.39	4.48	4.54	4.59	4.64	4.68	4.71	4.77	4.84	4.92	5.02	5.16	5.22	
120	.10	2.34	2.48	2.57	2.63	2.68	2.72	2.75	2.78	2.80	2.84	2.88	2.93	2.97	3.00	3.00	0.5
	.05	2.80	2.95	3.04	3.12	3.17	3.22	3.25	3.29	3.31	3.36	3.41	3.47	3.53	3.60	3.60	
	.01	3.70	3.86	3.96	4.04	4.11	4.16	4.20	4.24	4.27	4.33	4.39	4.47	4.57	4.70	4.77	
	.005	4.04	4.20	4.31	4.39	4.45	4.50	4.55	4.59	4.62	4.68	4.75	4.83	4.94	5.09	5.18	
∞	.10	2.33	2.46	2.55	2.62	2.67	2.71	2.75	2.78	2.80	2.84	2.89	2.95	3.02	3.11	3.16	0
	.05	2.77	2.92	3.02	3.09	3.15	3.19	3.23	3.26	3.29	3.34	3.40	3.47	3.55	3.67	3.73	
	.01	3.64	3.80	3.90	3.98	4.04	4.09	4.13	4.17	4.20	4.26	4.33	4.41	4.51	4.67	4.78	
	.005	3.97	4.12	4.22	4.30	4.36	4.42	4.46	4.50	4.53	4.59	4.66	4.74	4.85	5.02	5.14	
								$60/p$	6	5	4	3	2	1	0.6		

Duncan's multiple range test: an example. The yields of 7 varieties of barley grain were measured, the measurements being replicated 6 times. The varietal means, arranged in order of magnitude, are set out below:

Variety	A	F	G	D	C	B	E
Rank i	1	2	3	4	5	6	7
Mean yield m_i	49.6	58.1	61.0	61.5	67.6	71.2	71.3

A two-factor analysis of variance, using the F-test, revealed significant differences between these means. Duncan's multiple range test is used to analyse these differences in detail.

First we calculate the standard error (S.E.) of the means. The residual mean sum of squares was found to be $s^2 = 79.64$; it is based on $\nu = (7-1)(6-1) = 30$ degrees of freedom. The S.E. of the means is $\sqrt{(79.64/6)} = 3.643$.

From the table we extract factors D for $\nu = 30$ and $\alpha = 0.05$ say. These values are then multiplied by the S.E. of the means to give the critical ranges R_p:

p	2	3	4	5	6	7
D_p	2.89	3.03	3.13	3.20	3.25	3.29
R_p	10.5	11.0	11.4	11.7	11.8	12.0

If the range of a set of p adjacent means is greater than R_p, the differences between the means are judged to be significant at significance level α, with the proviso that the test is not to be applied to a subset of a set that is judged to contain no significant differences. In applying the test it is important to use a systematic procedure.

Starting from the largest mean m_7, we consider in turn the differences $m_7 - m_1$, $m_7 - m_2$, $m_7 - m_3$, ..., comparing each with the appropriate critical range:

$$m_7 - m_1 = 21.7 > R_7, \quad m_7 - m_2 = 13.2 > R_6, \quad m_7 - m_3 = 10.3 < R_5$$

The difference $m_7 - m_3$ is not significant; this is indicated by underlining the set m_3 to m_7.

Next, starting from m_6, we consider $m_6 - m_1$ and $m_6 - m_2$ (but not, in accordance with the proviso, $m_6 - m_3$ etc.):

$$m_6 - m_1 = 21.6 > R_6, \quad m_6 - m_2 = 13.1 > R_5; \text{ both are significant}$$

Next we start from m_5 and consider $m_5 - m_1$ and $m_5 - m_2$:

$$m_5 - m_1 = 18.0 > R_5, \quad m_5 - m_2 = 9.5 < R_4$$

The difference $m_5 - m_2$ is not significant; we underline the set m_2 to m_5.

Continuing in this manner we find:

$$m_4 - m_1 = 11.9 > R_4, \quad m_3 - m_1 = 11.4 > R_3, \quad m_2 - m_1 = 8.5 < R_2$$

The difference $m_2 - m_1$ is not significant; we underline the set m_1, m_2.

The results of the analysis are embodied in the first table. There are no significant differences between the means in any of the three underlined sets at significance level 0.05, but there *are* significant differences between means that do not all belong to any one such set.

TESTS BASED ON t-, CHI-SQUARE, F- AND HYPERGEOMETRIC DISTRIBUTIONS

Estimation of population mean and variance. The mean $\bar{x}$ of a sample $x_1, x_2, \ldots, x_n$ is $\Sigma_i x_i / n$. For random samples of size n taken from a normal population with mean μ and variance σ^2, $\bar{x}$ has a normal distribution with mean μ and variance $\sigma_{\bar{x}}^2 = \sigma^2/n$, so that $(\bar{x} - \mu)/\sigma_{\bar{x}} \sim \mathrm{N}(0, 1)$. The statistic $\bar{x}$ is a good estimator of μ; by using the quantiles tabulated on p. 30, confidence intervals for μ with bounds $\bar{x} - z_{[P]}\sigma_{\bar{x}}$ can be formed.

The sample variance s^2 is defined as $\Sigma_i (x_i - \bar{x})^2/(n-1)$. If σ^2 is unknown we use $\hat{\sigma}^2 = s^2$ as an estimator, so that in place of $\sigma_{\bar{x}}^2$ we have $\hat{\sigma}_{\bar{x}}^2 = \hat{\sigma}^2/n = s^2/n$. The sampling distribution of $(\bar{x} - \mu)/\hat{\sigma}_{\bar{x}}$ or $\sqrt{n}(\bar{x} - \mu)/s$ is, by definition, the t-distribution with $\nu = n-1$ degrees of freedom. We write $(\bar{x} - \mu)/\hat{\sigma}_{\bar{x}} \sim t(n-1)$. By using quantiles of this distribution, confidence intervals for μ can be formed.

Equivalently we may use the fact that $(\bar{x} - \mu)^2/\hat{\sigma}_{\bar{x}}^2$ is the ratio of two independent chi-square variables and accordingly has the distribution $F(1, \nu)$; quantiles are tabulated on pp. 40-3.

Estimation of difference between population means. If independent random samples $x_1, x_2, \ldots, x_m$ and $y_1, y_2, \ldots, y_n$ are taken from two normal populations with means μ_x and μ_y and equal variances σ^2, the difference $\bar{d} = \bar{x} - \bar{y}$ has a normal sampling distribution with mean $\mu_x - \mu_y$ and variance $\sigma_{\bar{d}}^2 = \sigma^2(1/m + 1/n)$. It is a good estimator of $\mu_x - \mu_y$.

If σ^2 also is unknown we estimate it by $\hat{\sigma}^2$, formed by pooling the sums of squares of the two samples:

$$\hat{\sigma}^2 = \{\sum_i (x_i - \bar{x})^2 + \sum_i (y_i - \bar{y})^2\}/(m+n-2) = \{(m-1)s_x^2 + (n-1)s_y^2\}/(m+n-2)$$

For $\sigma_{\bar{d}}^2$ we use the estimator $\hat{\sigma}_{\bar{d}}^2 = \hat{\sigma}^2(1/m + 1/n)$. The distribution of $\{(\bar{x} - \bar{y}) - (\mu_x - \mu_y)\}/\hat{\sigma}_{\bar{d}}$ is $t(m+n-2)$. By using the tabulated quantiles of this distribution (p. 39) we can construct confidence intervals for $\mu_x - \mu_y$, and test the hypothesis that $\mu_x - \mu_y$ has a postulated value.

Paired comparison test. If we have a random sample of matched pairs (x_i, y_i), we can use a t-test to compare the means of the x- and y-populations. The procedure follows that used for estimating the mean of a single population, but with the difference score $d_i = x_i - y_i$ (assumed to be normally distributed) in place of x_i.

Confidence intervals for population variance. If a random sample $z_1, z_2, \ldots, z_n$ is taken from a standard normal population, the sampling distribution of $\Sigma_i z_i^2$ is, by definition, the chi-square distribution with n degrees of freedom, $\chi^2(n)$. Hence, if $x_1, x_2, \ldots, x_n$ is a random sample from a normal distribution with mean μ and variance σ^2, $\Sigma_i (x_i - \mu)^2/\sigma^2 \sim \chi^2(n)$. The distribution of $\Sigma_i (x_i - \bar{x})^2/\sigma^2$ or $(n-1)s^2/\sigma^2$ is $\chi^2(n-1)$. This fact can be used to construct confidence intervals for σ^2.

For example, suppose that, for a sample of 10 observations from a normal distribution, $\Sigma x = 49$ and $\Sigma x^2 = 253$. Then $\Sigma(x-\bar{x})^2 = 253 - 49^2/10 = 12.9$, and $s^2 = 12.9/9 = 1.43$. Thus a point estimate of σ^2 is 1.43. To obtain bounds of a 0.95 confidence interval for σ^2 we divide the sum of squares by the 0.025 and 0.975 quantiles (i.e. $2\frac{1}{2}\%$ and $97\frac{1}{2}\%$ percentiles) of $\chi^2(9)$, giving $12.9/2.700 = 4.78$ and $12.9/19.02 = 0.68$.

Variance ratio tests. If two random samples of sizes n_1 and n_2, taken from normal populations with variances σ_1^2 and σ_2^2 respectively, have variances s_1^2 and s_2^2, then

$$\left(\frac{s_1^2}{\sigma_1^2}\right)\Big/\left(\frac{s_2^2}{\sigma_2^2}\right)$$

has the distribution $F(n_1-1, n_2-1)$; see p. 43. This fact can be used to construct significance tests and confidence intervals for the variance ratio σ_1^2/σ_2^2.

For example, suppose that $n_1 = 16$, $n_2 = 10$, $s_1^2 = 172.36$, $s_2^2 = 80.30$, and that we wish to test the null hypothesis $H_0 : \sigma_1^2 = \sigma_2^2$ against the two-sided alternative hypothesis $H_1 : \sigma_1^2 \neq \sigma_2^2$ at significance level 0.1. Under $H_0, s_1^2/s_2^2$ has the distribution $F(15, 9)$ and we reject H_0 if the variance ratio is less than $F_{[.05]}$ or greater than $F_{[.95]}$. Now, from p. 41,

$$F_{[.95]}(15, 9) = 3.01 \quad F_{[.05]}(15, 9) = 1/F_{[.95]}(9, 15) = 1/2.59 = 0.386$$

The variance ratio is $172.36/80.30$ or 2.146, and lies between the two quantiles, so we cannot reject the null hypothesis at significance level 0.1. To construct bounds of a 0.9 confidence interval for σ_1^2/σ_2^2 we divide the variance ratio by the two quantiles, thus:

$$2.146/3.01 = 0.71 \quad 2.146/0.386 = 5.56$$

Tests for goodness of fit. These tests are used to decide whether an observed frequency distribution is compatible with some hypothesised or theoretical frequency distribution. Suppose we have N observations each of which is placed in one of k mutually exclusive classes. We use as test statistic

$$U = \sum_i (f_i - \tilde{f}_i)^2/\tilde{f}_i \quad \text{or equivalently} \quad \sum_i f_i^2/\tilde{f}_i - N$$

where f_i is the frequency of observations falling in class i and $\tilde{f}_i$ is the expected frequency according to the hypothesised distribution. The sampling distribution of U is approximately $\chi^2(k-r-1)$ where r is the number of independent parameters (apart from N) estimated to fit the sample data. The approximation is reasonably good if the expected frequencies are not too small; as a working rule none of them should be less than 1 and not more than one-fifth should be less than 5. If necessary, classes should be merged.

Alternatively we can use a G-test. If, according to the hypothesised distribution, the probability that an observation falls in class i is $\tilde{p}_i$ (so that $\tilde{f}_i = N\tilde{p}_i$), the probability of occurrence of the observed frequencies $f_1, f_2, \ldots$ is given by a term in the multinomial expansion of $(\tilde{p}_1 + \tilde{p}_2 + \ldots + \tilde{p}_k)^N$, namely:

$$\frac{N!}{f_1! f_2! \ldots} \tilde{p}_1^{f_1} \tilde{p}_2^{f_2} \ldots$$

(This is a *multinomial distribution*.) For the distribution in which the class probabilities are those estimated from the observed frequencies (namely $\hat{p}_i = f_i/N$) the probability of the observed frequencies is given by a similar expression with f_i/N in place of $\tilde{p}_i$. The ratio of the two probabilities (the *likelihood ratio*) is:

$$L = \prod_i \left(\frac{\tilde{p}_i}{f_i/N}\right)^{f_i} = \prod_i \left(\frac{\tilde{f}_i}{f_i}\right)^{f_i}$$

As test statistic we use:

$$G = -2 \ln L = 2\{\Sigma f_i \ln f_i - \Sigma f_i \ln \tilde{p}_i - N \ln N\} = 2\{\Sigma f_i \ln f_i - \Sigma f_i \ln \tilde{f}_i\}$$

Both of these expressions are useful. For large samples the G-distribution like the U-distribution approximates to $\chi^2(k-r-1)$, where r is the number of independent parameters (apart from N) estimated to fit the sample data.

Tests of independence in contingency tables

Tests of independence of two criteria of classification. Suppose that a random sample of N individuals is divided into mutually exclusive classes 1 to n by criterion 1 (e.g. by race) and into mutually exclusive classes 1 to k by criterion 11 (e.g. by blood group). We denote by f_{ij} the number of individuals placed in class i by 1 and class j by 11: these frequencies are set out in a two-way *contingency table*; the as and bs are row and column totals. If the criteria were independent the expected frequencies in each row would be proportional. If we estimated them from the observed marginal frequencies they would be:

$$\hat{f}_{ij} = a_i b_j / N$$

	II				
I	1	2	$\ldots$ k	
1	f_{11}	f_{12}	$\ldots$ f_{1k}	a_1
2	f_{21}	f_{22}	$\ldots$ f_{2k}	a_2
$\vdots$	$\vdots$	$\vdots$	$\vdots$	$\vdots$
n	f_{n1}	f_{n2}	$\ldots$ f_{nk}	a_n
	b_1	b_2	$\ldots$ b_k	N

As test statistic we use:

$$U = \sum_i \sum_j \frac{(f_{ij} - \hat{f}_{ij})^2}{\hat{f}_{ij}} \quad \text{or} \quad \sum_i \sum_j \frac{f_{ij}^2}{\hat{f}_{ij}} - N$$

(The second form is more convenient for calculation.) For large samples the distribution of U is approximately chi-square with $(k-1)(n-1)$ degrees of freedom; $U \sim \chi^2((k-1)(n-1))$.

Alternatively a G-test may be employed. The probability distribution is a multinomial distribution. As test statistic we use the log likelihood ratio G (see the goodness of fit test described above):

$$G = -2 \ln \prod_i \prod_j \left(\frac{\hat{f}_{ij}}{f_{ij}}\right)^{f_{ij}} = 2\left\{ \sum_i \sum_j f_{ij} \ln f_{ij} - \sum_i a_i \ln a_i - \sum_j b_j \ln b_j + N \ln N \right\}$$

For large N, $G \sim \chi^2((k-1)(n-1))$ approximately.

For a 2×2 contingency table the statistic U takes the form:

$$U = \frac{(f_{11} f_{22} - f_{12} f_{21})^2 N}{a_1 a_2 b_1 b_2}$$

The distribution of U is approximately $\chi^2(1)$ if none of the expected frequencies is less than 5.

Tests of homogeneity. These are used to test the homogeneity of two or more populations with respect to some criterion of classification. We suppose that a random sample of observations is available from each of n populations, and that each observation can be placed in one of k mutually exclusive classes. We denote by f_{ij} the frequency of observations from sample i falling in class j; these frequencies are set out in a two-way table. The row total a_i is now the size of sample i; b_j is the total number of observations falling in class j; N is the total number of observations. If the populations are homogeneous with respect to the classification, the expected frequencies in each row must be proportional, and can be expressed in terms of the marginal frequencies in the same way as for the preceding test. Test statistics U and G are formed and used in the same way as in that test.

	Class			
	1	2	$\ldots$ k	
1	f_{11}	f_{12}	$\ldots$ f_{1k}	a_1
2	f_{21}	f_{22}	$\ldots$ f_{2k}	a_2
$\vdots$	$\vdots$	$\vdots$	$\vdots$	$\vdots$
n	f_{n1}	f_{n2}	$\ldots$ f_{nk}	a_n
	b_1	b_2	$\ldots$ b_k	N

Median tests. These are used to test for significant differences between the medians of two or more populations. Like the sign test they can be used when the t-test is inapplicable. This is a particular case of the tests of homogeneity described above, the observations being separated into two classes according to the criterion $x \gtrless M$, where M is the median of the pooled observations. Observations exactly on the median are discarded. Note that all the marginal totals in the frequency table are now fixed; the as are sample sizes and $b_1 = b_2 = \frac{1}{2}N$. If we have just two populations and the samples are small, Fisher's 'exact' test could be used instead of a chi-square test.

Westenberg's interquartile range test. This is a test for differences between the dispersion of two or more populations. It is similar to the median test, but the two column classes are now defined by $Q_L < x < Q_U$ and $(x < Q_L$ or $x > Q_U)$ where Q_L and Q_U are the lower and upper quartiles of the pooled observations. This is not strictly a test of equality of interquartile ranges, but rather of equality of the proportions of population values falling within the interquartile range of the pooled observations.

Fisher's exact test. In some situations all the marginal totals in a contingency table are fixed. For example, the following problem arises in the context of sampling inspection. A batch of N items contains k defective items; k is unknown. A sample of n items taken from the batch is found to contain r defective items. The problem is to estimate k. As a first step we specify the probability distribution of r for fixed k. The data are set out in a contingency table, where now all the marginal totals are fixed. The probability that exactly r defective items are found in the sample is:

	Defective	Perfect	
Sample	r	$n-r$	n
Remainder	$k-r$	$N-k-n+r$	$N-n$
	k	$N-k$	N

$$p(r) = \binom{k}{r}\binom{N-k}{n-r}\bigg/\binom{N}{n} = \frac{k!\,(N-k)!\,n!\,(N-n)!}{r!\,(k-r)!\,(n-r)!\,(N-k-n+r)!\,N!}$$

This defines a *hypergeometric distribution*. A test of a 2×2 contingency table with fixed marginal totals, based on this distribution, is called *Fisher's exact test*.

If we revert to the notation used earlier, the probability distribution is:

$$p(x) = \frac{a_1!\,a_2!\,b_1!\,b_2!}{f_{11}!\,f_{12}!\,f_{21}!\,f_{22}!\,N!}$$

f_{11}	f_{12}	a_1
f_{21}	f_{22}	a_2
b_1	b_2	N

where x is any one of the f_{ij}; the remaining f_{ij} are determined by x and the marginal totals.

For the tables on pp. 52–57, x has been taken to be f_{21}. The tables give lower tail critical values for the distribution of f_{21} for fixed marginal totals. These critical values are however arranged according to the values of a_1, a_2 and f_{11}. They cover all contingency tables for which either both row totals or both column totals are $\leqslant 15$.

Before we use the tables the following preliminary steps should, if necessary, be taken in the order given:

(i) interchange rows and columns so that both *row* totals become $\leqslant 15$,
(ii) interchange the rows to make $a_1 \geqslant a_2$,
(iii) interchange the columns to make $f_{11}/a_1 \geqslant f_{21}/a_2$ (or equivalently $f_{11}f_{22} \geqslant f_{12}f_{21}$).

The tables then give critical values $f_{21(\alpha)}$ for significance levels $\alpha = 0.05, 0.025, 0.01$ and 0.005.

The null hypothesis (under which the null distribution of f_{21} is a hypergeometric distribution) is rejected at significance level α if $\qquad (f_{21})_{\text{observed}} \leqslant f_{21(\alpha)}$

so that:

$$\text{Prob}\left((f_{11}/a_1 - f_{21}/a_2) \geqslant (f_{11}/a_1 - f_{21}/a_2)_{\text{observed}}\right) < \alpha$$

Fisher's test is thus a one-sided test of the inequality of the proportions in the two rows of the contingency table. The corresponding chi-square tests (using U or G as test statistic) are two-sided.

Example. 13 boys and 15 girls from the same age group, selected randomly, took part in an essay-writing competition. Six prizes were awarded for the best essays. The results are shown in table (a) (B boys, G girls, S successful, U unsuccessful). The boys' results were better than the girls'. Were they *significantly* better at level 0.05? We use Fisher's test. Rearrangement of the contingency table in accordance with rules (i)–(iii) gives table (b). Then we find on p. 56 that for $a_1 = 15$, $a_2 = 13$ and $f_{11} = 14$ the critical value of f_{21} at significance level 0.05 is 7. Since the observed value is 8, we conclude that the difference between the proportion of successes is not significant at level 0.05.

(a)	S	U	
B	5	8	13
G	1	14	15
	6	22	28

(b)	U	S	
G	14	1	15
B	8	5	13
	22	6	28

Tests of association. Given a random sample of observations from a bivariate population, a chi-square test or Fisher's test can be used to test for association between the variables X and Y.

Example 1. The scatter diagram (fig. 12) represents a sample of 24 observations. We divide it into four quarters by drawing the lines $x = M_X$ and $y = M_Y$, where M_X and M_Y are the medians of the values of X and Y respectively. The numbers of points in the four quarters are set out as a contingency table (*a*), points on the median lines being discarded. We use Fisher's test. The table, rearranged in accordance with rules (i)–(iii), appears as (*b*). The table on p. 54 then shows that for $a_1 = a_2 = 12$ and $f_{11} = 8$ the critical value of f_{21} is 3 at significance level 0.05. Since the observed value is 4 the association is not significant at this level.

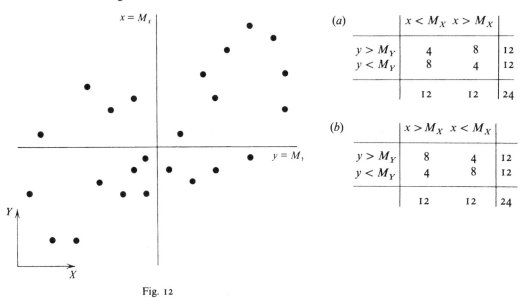

(*a*)

	$x < M_X$	$x > M_X$	
$y > M_Y$	4	8	12
$y < M_Y$	8	4	12
	12	12	24

(*b*)

	$x > M_X$	$x < M_X$	
$y > M_Y$	8	4	12
$y < M_Y$	4	8	12
	12	12	24

Fig. 12

Example 2. The scatter diagram below represents a sample of 30 observations. We wish to test the null hypothesis H_0 of no association between X and Y by a test sensitive to both linear and non-linear association. Accordingly we divide the points into three 'columns' containing 10 points each and two

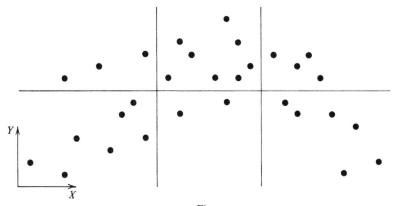

Fig. 13

'rows' containing 15 points each. Under H_0 the expected number in each cell is 5. Hence:

$$U = (2^2+3^2+1^2+2^2+3^2+1^2)/5 = 28/5 = 5.6$$

	C_1	C_2	C_3	
R_1	3	8	4	15
R_2	7	2	6	15
	10	10	10	30

Since U lies between the quantiles $w_{[.90]}$ and $w_{[.95]}$ of $\chi^2(2)$ we can reject H_0 at significance level 0.1 but not at 0.05.

Alternatively we can combine columns C_1 and C_3 to form a new column C'_1. We then have the 2×2 table (a) and can use Fisher's test. This is now a one-sided test of H_0 against the alternative hypothesis H_1 that Y tends to take higher values in the middle of the range of values of X than at the extremes. Rearrangement in accordance with rules (i)–(iii) gives table (b). Then, from the table on p. 56, we have that for $a_1 = a_2 = 15$ and $f_{11} = 8$ the critical value of f_{21} at significance level 0.05 is 2. Since the observed value of f_{21} is 2 we can reject H_0 in favour of H_1 at significance level 0.05.

(a)

	C'_1	C_2	
R_1	7	8	15
R_2	13	2	15
	20	10	30

(b)

	C_2	C'_1	
R_1	8	7	15
R_2	2	13	15
	10	20	30

The Cochran Q test. Suppose that n individuals (or groups) are subjected to k treatments, and that the responses (e.g. success or failure) are scored 1 or 0. These scores are set out in an $n \times k$ table; the row total a_i is the total score of individual (or group) i; the column total b_j is the total score under treatment j. In order to test the null hypothesis that the treatments are equally effective (so that the probability of success is the same under all treatments) we use as test statistic

$$Q = \frac{k(k-1)\Sigma_j b_j^2 - (k-1)N^2}{\Sigma_i(k-a_i)a_i}$$

where N is the grand total score. The distribution of Q is approximately $\chi^2(k-1)$ for large n.

If, for $k = 2$, we take columns 1 and 2 to represent responses before and after treatment, the test becomes the McNemar test for the significance of changes; Q becomes the statistic U used for that test.

Example. Twelve children were given a test of physical fitness. The test was repeated twice at intervals of 3 months. The results are shown in the table, where 1 indicates success and 0 indicates failure. At what significance level are the differences between the results of the three tests significant? The Q-test is appropriate. Since $k = 3$:

$$Q = \frac{3 \times 2 \times (5^2 + 10^2 + 8^2) - 2 \times 23^2}{3 \times 2 + 4 \times 2}$$

$$= \frac{6 \times 189 - 2 \times 529}{6+8}$$

$$= 5.43$$

The distribution of Q is approximately $\chi^2(2)$, for which:

$$w_{[.90]} = 4.60 \quad w_{[.95]} = 5.99$$

Since the observed value of Q lies between these quantiles, the differences between the results of the three tests is significant at level 0.1 but not at 0.05.

	Test			
Child	I	2	3	
I	0	I	I	2
2	I	I	I	3
3	0	I	0	I
4	0	I	I	2
5	0	0	0	0
6	I	I	I	3
7	0	I	I	2
8	0	I	I	2
9	I	0	0	I
10	0	I	0	I
11	I	I	I	3
12	I	I	I	3
	5	10	8	23

LOWER CRITICAL VALUES OF f_{21} FOR FISHER'S EXACT TEST

a_1	a_2	f_{11}	Significance level .05	.025	.01	.005
3	3	3	0	–	–	–
4	4	4	0	0	–	–
	3	4	0	–	–	–
5	5	5	1	1	0	0
		4	0	0	–	–
	4	5	1	0	0	–
		4	0	–	–	–
	3	5	0	0	–	–
	2	5	0	–	–	–
6	6	6	2	1	1	0
		5	1	0	0	–
		4	0	–	–	–
	5	6	1	1	0	0
		5	0	0	–	–
		4	0	–	–	–
	4	6	1	0	0	0
		5	0	0	–	–
	3	6	0	0	–	–
		5	0	–	–	–
	2	6	0	–	–	–
7	7	7	3	2	1	1
		6	1	1	0	0
		5	0	0	–	–
		4	0	–	–	–
	6	7	2	2	1	1
		6	1	0	0	0
		5	0	0	–	–
		4	0	–	–	–
	5	7	2	1	0	0
		6	1	0	0	–
		5	0	–	–	–
	4	7	1	1	0	0
		6	0	0	–	–
		5	0	–	–	–
	3	7	0	0	0	–
		6	0	–	–	–
	2	7	0	–	–	–
8	8	8	4	3	2	2
		7	2	2	1	0
		6	1	1	0	0
		5	0	0	–	–
		4	0	–	–	–
	7	8	3	2	2	1
		7	2	1	1	0
		6	1	0	0	–
		5	0	0	–	–
	6	8	2	2	1	1
		7	1	1	0	0
		6	0	0	0	–
		5	0	–	–	–

a_1	a_2	f_{11}	Significance level .05	.025	.01	.005
8	5	8	2	1	1	0
		7	1	0	0	0
		6	0	0	–	–
		5	0	–	–	–
	4	8	1	1	0	0
		7	0	0	–	–
		6	0	–	–	–
	3	8	0	0	0	–
		7	0	0	–	–
	2	8	0	0	–	–
9	9	9	5	4	3	3
		8	3	3	2	1
		7	2	1	1	0
		6	1	1	0	0
		5	0	0	–	–
		4	0	–	–	–
	8	9	4	3	3	2
		8	3	2	1	1
		7	2	1	0	0
		6	1	0	0	–
		5	0	0	–	–
	7	9	3	3	2	2
		8	2	2	1	0
		7	1	1	0	0
		6	0	0	–	–
		5	0	–	–	–
	6	9	3	2	1	1
		8	2	1	0	0
		7	1	0	0	–
		6	0	0	–	–
		5	0	–	–	–
	5	9	2	1	1	1
		8	1	1	0	0
		7	0	0	–	–
		6	0	–	–	–
	4	9	1	1	0	0
		8	0	0	0	–
		7	0	0	–	–
		6	0	–	–	–
	3	9	1	0	0	0
		8	0	0	–	–
		7	0	–	–	–
	2	9	0	0	–	–
10	10	10	6	5	4	3
		9	4	3	3	2
		8	3	2	1	1
		7	2	1	1	0
		6	1	0	0	–
		5	0	0	–	–
		4	0	–	–	–

The table is applicable only if:

$$a_1 \geqslant a_2$$

$$f_{11}/a_1 \geqslant f_{21}/a_2 \quad \text{or equivalently} \quad f_{11}f_{22} \geqslant f_{12}f_{21}$$

f_{11}	f_{12}	a_1
f_{21}	f_{22}	a_2
b_1	b_2	N

LOWER CRITICAL VALUES OF f_{21} FOR FISHER'S EXACT TEST

a_1	a_2	f_{11}	.05	.025	.01	.005
10	9	10	5	4	3	3
		9	4	3	2	2
		8	2	2	1	1
		7	1	1	0	0
		6	1	0	0	–
		5	0	0	–	–
	8	10	4	4	3	2
		9	3	2	2	1
		8	2	1	1	0
		7	1	1	0	0
		6	0	0	–	–
		5	0	–	–	–
	7	10	3	3	2	2
		9	2	2	1	1
		8	1	1	0	0
		7	1	0	0	–
		6	0	0	–	–
		5	0	–	–	–
	6	10	3	2	2	1
		9	2	1	1	0
		8	1	1	0	0
		7	0	0	–	–
		6	0	–	–	–
	5	10	2	2	1	1
		9	1	1	0	0
		8	1	0	0	–
		7	0	0	–	–
		6	0	–	–	–
	4	10	1	1	0	0
		9	1	0	0	0
		8	0	0	–	–
		7	0	–	–	–
	3	10	1	0	0	0
		9	0	0	–	–
		8	0	–	–	–
	2	10	0	0	–	–
		9	0	–	–	–
11	11	11	7	6	5	4
		10	5	4	3	3
		9	4	3	2	2
		8	3	2	1	1
		7	2	1	0	0
		6	1	0	0	–
		5	0	0	–	–
		4	0	–	–	–

a_1	a_2	f_{11}	.05	.025	.01	.005
11	10	11	6	5	4	4
		10	4	4	3	2
		9	3	3	2	1
		8	2	2	1	0
		7	1	1	0	0
		6	1	0	0	–
		5	0	0	–	–
	9	11	5	4	4	3
		10	4	3	2	2
		9	3	2	1	1
		8	2	1	1	0
		7	1	1	0	0
		6	0	0	–	–
		5	0	–	–	–
	8	11	4	4	3	3
		10	3	3	2	1
		9	2	2	1	1
		8	1	1	0	0
		7	1	0	0	–
		6	0	0	–	–
		5	0	–	–	–
	7	11	4	3	2	2
		10	3	2	1	1
		9	2	1	1	0
		8	1	1	0	0
		7	0	0	–	–
		6	0	0	–	–
	6	11	3	2	2	1
		10	2	1	1	0
		9	1	1	0	0
		8	1	0	0	–
		7	0	0	–	–
		6	0	–	–	–
	5	11	2	2	1	1
		10	1	1	0	0
		9	1	0	0	0
		8	0	0	–	–
		7	0	–	–	–
	4	11	1	1	1	0
		10	1	0	0	0
		9	0	0	–	–
		8	0	–	–	–
	3	11	1	0	0	0
		10	0	0	–	–
		9	0	–	–	–
	2	11	0	0	–	–
		10	0	–	–	–

For given values of a_1, a_2 and f_{11}, the proportion f_{11}/a_1 is significantly larger than f_{21}/a_2 if the observed value of f_{21} does not exceed the tabulated critical value. Where the entry is –, or there is no entry, there is no significant 2×2 table.

LOWER CRITICAL VALUES OF f_{21} FOR FISHER'S EXACT TEST

a_1	a_2	f_{11}	Significance level				a_1	a_2	f_{11}	Significance level			
			.05	.025	.01	.005				.05	.025	.01	.005
12	12	12	8	7	6	5	12	6	12	3	3	2	2
		11	6	5	4	4			11	2	2	1	1
		10	5	4	3	2			10	1	1	0	0
		9	4	3	2	1			9	1	0	0	0
		8	3	2	1	1			8	0	0	-	-
		7	2	1	0	0			7	0	0	-	-
		6	1	0	0	-			6	0	-	-	-
		5	0	0	-	-		5	12	2	2	1	1
		4	0	-	-	-			11	1	1	1	0
	11	12	7	6	5	5			10	1	0	0	0
		11	5	5	4	3			9	0	0	0	-
		10	4	3	2	2			8	0	0	-	-
		9	3	2	2	1			7	0	-	-	-
		8	2	1	1	0		4	12	2	1	1	0
		7	1	1	0	0			11	1	0	0	0
		6	1	0	0	-			10	0	0	0	-
		5	0	0	-	-			9	0	0	-	-
	10	12	6	5	5	4			8	0	-	-	-
		11	5	4	3	3		3	12	1	0	0	0
		10	4	3	2	2			11	0	0	0	-
		9	3	2	1	1			10	0	0	-	-
		8	2	1	0	0			9	0	-	-	-
		7	1	0	0	0		2	12	0	0	-	-
		6	0	0	-	-			11	0	-	-	-
		5	0	-	-	-	13	13	13	9	8	7	6
	9	12	5	5	4	3			12	7	6	5	4
		11	4	3	3	2			11	6	5	4	3
		10	3	2	2	1			10	4	4	3	2
		9	2	2	1	0			9	3	3	2	1
		8	1	1	0	0			8	2	2	1	0
		7	1	0	0	-			7	2	1	0	0
		6	0	0	-	-			6	1	0	0	-
		5	0	-	-	-			5	0	0	-	-
	8	12	5	4	3	3			4	0	-	-	-
		11	3	3	2	2		12	13	8	7	6	5
		10	2	2	1	1			12	6	5	5	4
		9	2	1	1	0			11	5	4	3	3
		8	1	1	0	0			10	4	3	2	2
		7	0	0	-	-			9	3	2	1	1
		6	0	0	-	-			8	2	1	1	0
	7	12	4	3	3	2			7	1	1	0	0
		11	3	2	2	1			6	1	0	0	-
		10	2	1	1	0			5	0	0	-	-
		9	1	1	0	0		11	13	7	6	5	5
		8	1	0	0	-			12	6	5	4	3
		7	0	0	-	-			11	4	4	3	2
		6	0	-	-	-			10	3	3	2	1
									9	3	2	1	1
									8	2	1	0	0
									7	1	0	0	0
									6	0	0	-	-
									5	0	-	-	-

The table is applicable only if:

$$a_1 \geqslant a_2$$

$$f_{11}/a_1 \geqslant f_{21}/a_2 \quad \text{or equivalently} \quad f_{11}f_{22} \geqslant f_{12}f_{21}$$

f_{11}	f_{12}	a_1
f_{21}	f_{22}	a_2
b_1	b_2	N

LOWER CRITICAL VALUES OF f_{21} FOR FISHER'S EXACT TEST

a_1	a_2	f_{11}	.05	.025	.01	.005	a_1	a_2	f_{11}	.05	.025	.01	.005
13	10	13	6	6	5	4	13	4	13	2	1	1	0
		12	5	4	3	3			12	1	1	0	0
		11	4	3	2	2			11	0	0	0	–
		10	3	2	1	1			10	0	0	–	–
		9	2	1	1	0			9	0	–	–	–
		8	1	1	0	0		3	13	1	1	0	0
		7	1	0	0	–			12	0	0	0	–
		6	0	0	–	–			11	0	0	–	–
		5	0	–	–	–			10	0	–	–	–
	9	13	5	5	4	4		2	13	0	0	0	–
		12	4	4	3	2			12	0	–	–	–
		11	3	3	2	1	14	14	14	10	9	8	7
		10	2	2	1	1			13	8	7	6	5
		9	2	1	0	0			12	6	6	5	4
		8	1	1	0	0			11	5	4	3	3
		7	0	0	–	–			10	4	3	2	2
		6	0	0	–	–			9	3	2	2	1
		5	0	–	–	–			8	2	2	1	0
	8	13	5	4	3	3			7	1	1	0	0
		12	4	3	2	2			6	1	0	0	–
		11	3	2	1	1			5	0	0	–	–
		10	2	1	1	0			4	0	–	–	–
		9	1	1	0	0		13	14	9	8	7	6
		8	1	0	0	–			13	7	6	5	5
		7	0	0	–	–			12	6	5	4	3
		6	0	–	–	–			11	5	4	3	2
	7	13	4	3	3	2			10	4	3	2	2
		12	3	2	2	1			9	3	2	1	1
		11	2	2	1	1			8	2	1	1	0
		10	1	1	0	0			7	1	1	0	0
		9	1	0	0	0			6	1	0	–	–
		8	0	0	–	–			5	0	0	–	–
		7	0	0	–	–		12	14	8	7	6	6
		6	0	–	–	–			13	6	6	5	4
	6	13	3	3	2	2			12	5	4	4	3
		12	2	2	1	1			11	4	3	3	2
		11	2	1	1	0			10	3	3	2	1
		10	1	1	0	0			9	2	2	1	1
		9	1	0	0	–			8	2	1	0	0
		8	0	0	–	–			7	1	0	0	–
		7	0	–	–	–			6	0	0	–	–
	5	13	2	2	1	1			5	0	–	–	–
		12	2	1	1	0		11	14	7	6	6	5
		11	1	1	0	0			13	6	5	4	4
		10	1	0	0	–			12	5	4	3	3
		9	0	0	–	–			11	4	3	2	2
		8	0	–	–	–			10	3	2	1	1
									9	2	1	1	0
									8	1	1	0	0
									7	1	0	0	–
									6	0	0	–	–
									5	0	–	–	–

For given values of a_1, a_2 and f_{11}, the proportion f_{11}/a_1 is significantly larger than f_{21}/a_2 if the observed value of f_{21} does not exceed the tabulated critical value. Where the entry is –, or there is no entry, there is no significant 2×2 table.

LOWER CRITICAL VALUES OF f_{21} FOR FISHER'S EXACT TEST

Left table:

a_1	a_2	f_{11}	.05	.025	.01	.005
14	10	14	6	6	5	4
		13	5	4	4	3
		12	4	3	3	2
		11	3	3	2	1
		10	2	2	1	1
		9	2	1	0	0
		8	1	1	0	0
		7	0	0	0	–
		6	0	0	–	–
		5	0	–	–	–
	9	14	6	5	4	4
		13	4	4	3	3
		12	3	3	2	2
		11	3	2	1	1
		10	2	1	1	0
		9	1	1	0	0
		8	1	0	0	–
		7	0	0	–	–
		6	0	–	–	–
	8	14	5	4	4	3
		13	4	3	2	2
		12	3	2	2	1
		11	2	2	1	1
		10	2	1	0	0
		9	1	0	0	0
		8	0	0	0	–
		7	0	0	–	–
		6	0	–	–	–
	7	14	4	3	3	2
		13	3	2	2	1
		12	2	2	1	1
		11	2	1	1	0
		10	1	1	0	0
		9	1	0	0	–
		8	0	0	–	–
		7	0	–	–	–
	6	14	3	3	2	2
		13	2	2	1	1
		12	2	1	1	0
		11	1	1	0	0
		10	1	0	0	–
		9	0	0	–	–
		8	0	0	–	–
		7	0	–	–	–
	5	14	2	2	1	1
		13	2	1	1	0
		12	1	1	0	0
		11	1	0	0	0
		10	0	0	–	–
		9	0	0	–	–
		8	0	–	–	–

Right table:

a_1	a_2	f_{11}	.05	.025	.01	.005
14	4	14	2	1	1	1
		13	1	1	0	0
		12	1	0	0	0
		11	0	0	–	–
		10	0	0	–	–
	3	14	1	1	0	0
		13	0	0	0	–
		12	0	0	–	–
		11	0	–	–	–
	2	14	0	0	0	–
		13	0	0	–	–
		12	0	–	–	–
15	15	15	11	10	9	8
		14	9	8	7	6
		13	7	6	5	5
		12	6	5	4	4
		11	5	4	3	3
		10	4	3	2	2
		9	3	2	1	1
		8	2	1	1	0
		7	1	1	0	0
		6	1	0	0	–
		5	0	0	–	–
		4	0	–	–	–
	14	15	10	9	8	7
		14	8	7	6	6
		13	7	6	5	4
		12	6	5	4	3
		11	5	4	3	2
		10	4	3	2	1
		9	3	2	1	1
		8	2	1	1	0
		7	1	1	0	0
		6	1	0	–	–
		5	0	–	–	–
	13	15	9	8	7	7
		14	7	7	6	5
		13	6	5	4	4
		12	5	4	3	3
		11	4	3	2	2
		10	3	2	2	1
		9	2	2	1	0
		8	2	1	0	0
		7	1	0	0	–
		6	0	0	–	–
		5	0	–	–	–

The table is applicable only if:

$$a_1 \geqslant a_2$$

$$f_{11}/a_1 \geqslant f_{21}/a_2 \quad \text{or equivalently} \quad f_{11}f_{22} \geqslant f_{12}f_{21}$$

f_{11}	f_{12}	a_1
f_{21}	f_{22}	a_2
b_1	b_2	N

LOWER CRITICAL VALUES OF f_{21} FOR FISHER'S EXACT TEST

a_1	a_2	f_{11}	Significance level			
			.05	.025	.01	.005
15	12	15	8	7	7	6
		14	7	6	5	4
		13	6	5	4	3
		12	5	4	3	2
		11	4	3	2	2
		10	3	2	1	1
		9	2	1	1	0
		8	1	1	0	0
		7	1	0	0	–
		6	0	0	–	–
		5	0	–	–	–
	11	15	7	7	6	5
		14	6	5	4	4
		13	5	4	3	3
		12	4	3	2	2
		11	3	2	2	1
		10	2	2	1	1
		9	2	1	0	0
		8	1	1	0	0
		7	1	0	0	–
		6	0	0	–	–
		5	0	–	–	–
	10	15	6	6	5	5
		14	5	5	4	3
		13	4	4	3	2
		12	3	3	2	2
		11	3	2	1	1
		10	2	1	1	0
		9	1	1	0	0
		8	1	0	0	–
		7	0	0	–	–
		6	0	–	–	–
	9	15	6	5	4	4
		14	5	4	3	3
		13	4	3	2	2
		12	3	2	2	1
		11	2	2	1	1
		10	2	1	0	0
		9	1	1	0	0
		8	1	0	0	–
		7	0	0	–	–
		6	0	–	–	–

a_1	a_2	f_{11}	Significance level			
			.05	.025	.01	.005
15	8	15	5	4	4	3
		14	4	3	3	2
		13	3	2	2	1
		12	2	2	1	1
		11	2	1	1	0
		10	1	1	0	0
		9	1	0	0	–
		8	0	0	–	–
		7	0	–	–	–
		6	0	–	–	–
	7	15	4	4	3	3
		14	3	3	2	2
		13	2	2	1	1
		12	2	1	1	0
		11	1	1	0	0
		10	1	0	0	0
		9	0	0	–	–
		8	0	0	–	–
		7	0	–	–	–
	6	15	3	3	2	2
		14	2	2	1	1
		13	2	1	1	0
		12	1	1	0	0
		11	1	0	0	0
		10	0	0	0	–
		9	0	0	–	–
		8	0	–	–	–
	5	15	2	2	2	1
		14	2	1	1	1
		13	1	1	0	0
		12	1	0	0	0
		11	0	0	0	–
		10	0	0	–	..
		9	0	–	–	–
	4	15	2	1	1	1
		14	1	1	0	0
		13	1	0	0	0
		12	0	0	0	–
		11	0	0	–	–
		10	0	–	–	–
	3	15	1	1	0	0
		14	0	0	0	0
		13	0	0	–	–
		12	0	0	–	–
		11	0	–	–	–
	2	15	0	0	0	–
		14	0	0	–	–
		13	0	–	–	–

Source: For the tables on pp. 52–7, D. J. Finney, *Biometrika* **35** (1948), 145–56. This is reproduced in *Biometrika Tables for Statisticians*, vol. I (3rd edn), Table 38.

For given values of a_1, a_2 and f_{11}, the proportion f_{11}/a_1 is significantly larger than f_{21}/a_2 if the observed value of f_{21} does not exceed the tabulated critical value. Where the entry is –, or there is no entry, there is no significant 2×2 table.

NATURAL LOGARITHMS

x	0	1	2	3	4	5	6	7	8	9
1.0	0.0000	0100	0198	0296	0392	0488	0583	0677	0770	0862
1.1	0.0953	1044	1133	1222	1310	1398	1484	1570	1655	1740
1.2	0.1823	1906	1989	2070	2151	2231	2311	2390	2469	2546
1.3	0.2624	2700	2776	2852	2927	3001	3075	3148	3221	3293
1.4	0.3365	3436	3507	3577	3646	3716	3784	3853	3920	3988
1.5	0.4055	4121	4187	4253	4318	4383	4447	4511	4574	4637
1.6	0.4700	4762	4824	4886	4947	5008	5068	5128	5188	5247
1.7	0.5306	5365	5423	5481	5539	5596	5653	5710	5766	5822
1.8	0.5878	5933	5988	6043	6098	6152	6206	6259	6313	6366
1.9	0.6419	6471	6523	6575	6627	6678	6729	6780	6831	6881
2.0	0.6931	6981	7031	7080	7129	7178	7227	7275	7324	7372
2.1	0.7419	7467	7514	7561	7608	7655	7701	7747	7793	7839
2.2	0.7885	7930	7975	8020	8065	8109	8154	8198	8242	8286
2.3	0.8329	8372	8416	8459	8502	8544	8587	8629	8671	8713
2.4	0.8755	8796	8838	8879	8920	8961	9002	9042	9083	9123
2.5	0.9163	9203	9243	9282	9322	9361	9400	9439	9478	9517
2.6	0.9555	9594	9632	9670	9708	9746	9783	9821	9858	9895
2.7	0.9933	9969	1.0006	0043	0080	0116	0152	0188	0225	0260
2.8	1.0296	0332	0367	0403	0438	0473	0508	0543	0578	0613
2.9	1.0647	0682	0716	0750	0784	0818	0852	0886	0919	0953
3.0	1.0986	1019	1053	1086	1119	1151	1184	1217	1249	1282
3.1	1.1314	1346	1378	1410	1442	1474	1506	1537	1569	1600
3.2	1.1632	1663	1694	1725	1756	1787	1817	1848	1878	1909
3.3	1.1939	1969	2000	2030	2060	2090	2119	2149	2179	2208
3.4	1.2238	2267	2296	2326	2355	2384	2413	2442	2470	2499
3.5	1.2528	2556	2585	2613	2641	2669	2698	2726	2754	2782
3.6	1.2809	2837	2865	2892	2920	2947	2975	3002	3029	3056
3.7	1.3083	3110	3137	3164	3191	3218	3244	3271	3297	3324
3.8	1.3350	3376	3403	3429	3455	3481	3507	3533	3558	3584
3.9	1.3610	3635	3661	3686	3712	3737	3762	3788	3813	3838
4.0	1.3863	3888	3913	3938	3962	3987	4012	4036	4061	4085
4.1	1.4110	4134	4159	4183	4207	4231	4255	4279	4303	4327
4.2	1.4351	4375	4398	4422	4446	4469	4493	4516	4540	4563
4.3	1.4586	4609	4633	4656	4679	4702	4725	4748	4770	4793
4.4	1.4816	4839	4861	4884	4907	4929	4951	4974	4996	5019
4.5	1.5041	5063	5085	5107	5129	5151	5173	5195	5217	5239
4.6	1.5261	5282	5304	5326	5347	5369	5390	5412	5433	5454
4.7	1.5476	5497	5518	5539	5560	5581	5602	5623	5644	5665
4.8	1.5686	5707	5728	5748	5769	5790	5810	5831	5851	5872
4.9	1.5892	5913	5933	5953	5974	5994	6014	6034	6054	6074

Add proportional parts

1	2	3	4	5	6	7	8	9
10	20	30	40	49	59	69	79	89
10	19	29	38	48	57	67	76	86
9	18	28	37	46	55	64	74	83
9	18	27	36	45	54	63	72	81
9	17	26	35	44	52	61	70	78
8	17	25	34	42	50	59	67	76
8	16	25	33	41	49	57	66	74
8	16	24	32	40	48	56	64	72
8	16	23	31	39	47	55	62	70
8	15	23	30	38	45	53	60	68
7	15	22	29	36	44	51	58	66
7	14	21	28	35	42	49	56	63
7	14	20	27	34	41	48	54	61
7	13	20	26	33	40	46	53	59
6	13	19	25	32	38	44	50	57
6	12	19	25	31	37	43	50	56
6	12	18	24	30	36	42	48	54
6	12	17	23	29	35	41	46	52
6	11	17	22	28	34	39	45	50
5	11	16	22	27	32	38	43	49
5	10	15	20	26	31	36	41	46
5	10	15	20	25	29	34	39	44
5	10	14	19	24	29	34	38	43
5	9	14	19	23	28	33	38	42
4	9	13	18	22	26	31	35	40
4	9	13	17	21	26	30	34	39
4	8	12	16	20	25	29	33	37
4	8	12	16	20	23	27	31	35
4	8	11	15	19	23	27	30	34
4	7	11	15	19	22	26	30	33
4	7	11	14	18	22	25	29	32
4	7	11	14	18	21	25	28	32
3	7	10	14	17	20	24	27	31
3	7	10	13	16	20	23	26	30
3	6	10	13	16	19	22	26	29
3	6	9	12	15	19	22	25	28
3	6	9	12	15	18	21	24	27
3	6	9	12	15	17	20	23	26
3	6	8	11	14	17	20	22	25
3	5	8	11	14	16	19	22	24
3	5	8	11	13	16	19	22	24
3	5	8	10	13	16	18	21	23
3	5	8	10	13	15	18	20	23
2	5	7	10	12	15	17	20	22
2	5	7	10	12	14	17	19	22
2	5	7	10	12	14	17	19	22
2	5	7	9	12	14	16	18	21
2	4	7	9	11	13	15	18	20
2	4	7	9	11	13	15	18	20
2	4	7	9	11	13	15	18	20
2	4	6	8	11	13	15	17	19
2	4	6	8	10	13	15	17	19
2	4	6	8	10	12	14	16	18

Source: The tables on pp. 58, 59 are taken from *The Cambridge Elementary Mathematical Tables* (2nd edn), by J. C. P. Miller & F. C. Powell.

If $x = y \times 10^{\pm n}$ with $1 \leqslant y < 10$: $\ln x = \ln y + \ln 10^{\pm n} \approx \ln y \pm 2.30259\,n$

NATURAL LOGARITHMS

x	0	1	2	3	4	5	6	7	8	9	1	2	3	4	5	6	7	8	9
											Add proportional parts								
5.0	1.6094	6114	6134	6154	6174	6194	6214	6233	6253	6273	2	4	6	8	10	12	14	16	18
5.1	1.6292	6312	6332	6351	6371	6390	6409	6429	6448	6467	2	4	6	8	10	12	14	16	18
5.2	1.6487	6506	6525	6544	6563	6582	6601	6620	6639	6658	2	4	6	8	10	11	13	15	17
5.3	1.6677	6696	6715	6734	6752	6771	6790	6808	6827	6845	2	4	6	8	9	11	13	15	17
5.4	1.6864	6882	6901	6919	6938	6956	6974	6993	7011	7029	2	4	5	7	9	11	13	14	16
5.5	1.7047	7066	7084	7102	7120	7138	7156	7174	7192	7210	2	4	5	7	9	11	13	14	16
5.6	1.7228	7246	7263	7281	7299	7317	7334	7352	7370	7387	2	4	5	7	9	11	13	14	16
5.7	1.7405	7422	7440	7457	7475	7492	7509	7527	7544	7561	2	3	5	7	9	10	12	14	15
5.8	1.7579	7596	7613	7630	7647	7664	7681	7699	7716	7733	2	3	5	7	9	10	12	14	15
5.9	1.7750	7766	7783	7800	7817	7834	7851	7867	7884	7901	2	3	5	7	8	10	12	14	15
6.0	1.7918	7934	7951	7967	7984	8001	8017	8034	8050	8066	2	3	5	6	8	10	11	13	14
6.1	1.8083	8099	8116	8132	8148	8165	8181	8197	8213	8229	2	3	5	6	8	10	11	13	14
6.2	1.8245	8262	8278	8294	8310	8326	8342	8358	8374	8390	2	3	5	6	8	10	11	13	14
6.3	1.8405	8421	8437	8453	8469	8485	8500	8516	8532	8547	2	3	5	6	8	10	11	13	14
6.4	1.8563	8579	8594	8610	8625	8641	8656	8672	8687	8703	2	3	5	6	8	10	11	13	14
6.5	1.8718	8733	8749	8764	8779	8795	8810	8825	8840	8856	2	3	5	6	8	9	11	12	14
6.6	1.8871	8886	8901	8916	8931	8946	8961	8976	8991	9006	2	3	5	6	8	9	11	12	14
6.7	1.9021	9036	9051	9066	9081	9095	9110	9125	9140	9155	1	3	4	6	7	9	10	12	13
6.8	1.9169	9184	9199	9213	9228	9242	9257	9272	9286	9301	1	3	4	6	7	9	10	12	13
6.9	1.9315	9330	9344	9359	9373	9387	9402	9416	9430	9445	1	3	4	6	7	8	10	11	13
7.0	1.9459	9473	9488	9502	9516	9530	9544	9559	9573	9587	1	3	4	6	7	8	10	11	13
7.1	1.9601	9615	9629	9643	9657	9671	9685	9699	9713	9727	1	3	4	6	7	8	10	11	13
7.2	1.9741	9755	9769	9782	9796	9810	9824	9838	9851	9865	1	3	4	6	7	8	10	11	13
7.3	1.9879	9892	9906	9920	9933	9947	9961	9974	9988		1	3	4	6	7	8	10	11	13
										2.0001	1	3	4	6	7	8	10	11	13
7.4	2.0015	0028	0042	0055	0069	0082	0096	0109	0122	0136	1	3	4	5	7	8	9	10	12
7.5	2.0149	0162	0176	0189	0202	0215	0229	0242	0255	0268	1	3	4	5	7	8	9	10	12
7.6	2.0281	0295	0308	0321	0334	0347	0360	0373	0386	0399	1	3	4	5	7	8	9	10	12
7.7	2.0412	0425	0438	0451	0464	0477	0490	0503	0516	0528	1	3	4	5	6	8	9	10	12
7.8	2.0541	0554	0567	0580	0592	0605	0618	0631	0643	0656	1	3	4	5	6	8	9	10	12
7.9	2.0669	0681	0694	0707	0719	0732	0744	0757	0769	0782	1	2	4	5	6	7	8	10	11
8.0	2.0794	0807	0819	0832	0844	0857	0869	0882	0894	0906	1	2	4	5	6	7	8	10	11
8.1	2.0919	0931	0943	0956	0968	0980	0992	1005	1017	1029	1	2	4	5	6	7	8	10	11
8.2	2.1041	1054	1066	1078	1090	1102	1114	1126	1138	1150	1	2	4	5	6	7	8	10	11
8.3	2.1163	1175	1187	1199	1211	1223	1235	1247	1258	1270	1	2	4	5	6	7	8	10	11
8.4	2.1282	1294	1306	1318	1330	1342	1353	1365	1377	1389	1	2	4	5	6	7	8	10	11
8.5	2.1401	1412	1424	1436	1448	1459	1471	1483	1494	1506	1	2	4	5	6	7	8	10	11
8.6	2.1518	1529	1541	1552	1564	1576	1587	1599	1610	1622	1	2	4	5	6	7	8	10	11
8.7	2.1633	1645	1656	1668	1679	1691	1702	1713	1725	1736	1	2	4	5	6	7	8	9	10
8.8	2.1748	1759	1770	1782	1793	1804	1815	1827	1838	1849	1	2	3	4	6	7	8	9	10
8.9	2.1861	1872	1883	1894	1905	1917	1928	1939	1950	1961	1	2	3	4	6	7	8	9	10
9.0	2.1972	1983	1994	2006	2017	2028	2039	2050	2061	2072	1	2	3	4	6	7	8	9	10
9.1	2.2083	2094	2105	2116	2127	2138	2148	2159	2170	2181	1	2	3	4	5	7	8	9	10
9.2	2.2192	2203	2214	2225	2235	2246	2257	2268	2279	2289	1	2	3	4	5	7	8	9	10
9.3	2.2300	2311	2322	2332	2343	2354	2364	2375	2386	2396	1	2	3	4	5	7	8	9	10
9.4	2.2407	2418	2428	2439	2450	2460	2471	2481	2492	2502	1	2	3	4	5	7	8	9	10
9.5	2.2513	2523	2534	2544	2555	2565	2576	2586	2597	2607	1	2	3	4	5	6	7	8	9
9.6	2.2618	2628	2638	2649	2659	2670	2680	2690	2701	2711	1	2	3	4	5	6	7	8	9
9.7	2.2721	2732	2742	2752	2762	2773	2783	2793	2803	2814	1	2	3	4	5	6	7	8	9
9.8	2.2824	2834	2844	2854	2865	2875	2885	2895	2905	2915	1	2	3	4	5	6	7	8	9
9.9	2.2925	2935	2946	2956	2966	2976	2986	2996	3006	3016	1	2	3	4	5	6	7	8	9

n	1	2	3	4	5	6	7	8	9	10
$\ln 10^{n}$	2.3026	4.6052	6.9078	9.2103	11.5129	13.8155	16.1181	18.4207	20.7233	23.0259
$\ln 10^{-n}$	$\bar{3}.6974$	$\bar{5}.3948$	$\bar{7}.0922$	$\overline{10}.7897$	$\overline{12}.4871$	$\overline{14}.1845$	$\overline{17}.8819$	$\overline{19}.5793$	$\overline{21}.2767$	$\overline{24}.9741$

$f \ln f$ (0–299)

f	0	1	2	3	4	5	6	7	8	9	f
0	0.000	0.000	1.386	3.296	5.545	8.047	10.751	13.621	16.636	19.775	0
10	23.026	26.377	29.819	33.344	36.947	40.621	44.361	48.165	52.027	55.944	10
20	59.915	63.935	68.003	72.116	76.273	80.472	84.711	88.988	93.302	97.652	20
30	102.036	106.454	110.904	115.385	119.896	124.437	129.007	133.604	138.228	142.879	30
40	147.555	152.256	156.982	161.732	166.504	171.300	176.118	180.957	185.818	190.699	40
50	195.601	200.523	205.465	210.425	215.405	220.403	225.420	230.454	235.506	240.575	50
60	245.661	250.763	255.882	261.017	266.169	271.335	276.517	281.714	286.927	292.153	60
70	297.395	302.650	307.920	313.204	318.501	323.812	329.136	334.473	339.823	345.186	70
80	350.562	355.950	361.351	366.764	372.189	377.625	383.074	388.534	394.006	399.489	80
90	404.983	410.488	416.005	421.532	427.070	432.618	438.177	443.747	449.327	454.917	90
100	460.517	466.127	471.747	477.377	483.017	488.666	494.325	499.993	505.670	511.357	100
110	517.053	522.758	528.472	534.195	539.927	545.667	551.416	557.174	562.941	568.716	110
120	574.499	580.291	586.091	591.899	597.715	603.539	609.372	615.212	621.060	626.916	120
130	632.779	638.651	644.530	650.416	656.311	662.212	668.121	674.037	679.961	685.892	130
140	691.830	697.775	703.727	709.687	715.653	721.626	727.607	733.594	739.587	745.588	140
150	751.595	757.609	763.630	769.657	775.691	781.731	787.778	793.831	799.890	805.956	150
160	812.028	818.106	824.191	830.281	836.378	842.481	848.590	854.705	860.826	866.953	160
170	873.086	879.224	885.369	891.519	897.676	903.838	910.005	916.179	922.357	928.542	170
180	934.732	940.928	947.129	953.336	959.548	965.766	971.989	978.217	984.451	990.690	180
190	996.935	1003.184	1009.439	1015.699	1021.964	1028.235	1034.510	1040.791	1047.077	1053.368	190
200	1059.663	1065.964	1072.270	1078.581	1084.896	1091.217	1097.542	1103.873	1110.208	1116.548	200
210	1122.893	1129.242	1135.596	1141.955	1148.319	1154.687	1161.060	1167.438	1173.820	1180.207	210
220	1186.598	1192.994	1199.394	1205.799	1212.209	1218.623	1225.041	1231.464	1237.891	1244.322	220
230	1250.758	1257.198	1263.643	1270.092	1276.545	1283.003	1289.464	1295.930	1302.400	1308.875	230
240	1315.353	1321.836	1328.323	1334.814	1341.309	1347.808	1354.312	1360.819	1367.330	1373.846	240
250	1380.365	1386.889	1393.416	1399.948	1406.483	1413.022	1419.565	1426.113	1432.664	1439.218	250
260	1445.777	1452.340	1458.906	1465.477	1472.051	1478.628	1485.210	1491.795	1498.385	1504.977	260
270	1511.574	1518.174	1524.778	1531.386	1537.997	1544.612	1551.231	1557.853	1564.479	1571.108	270
280	1577.741	1584.378	1591.018	1597.661	1604.309	1610.959	1617.614	1624.271	1630.933	1637.597	280
290	1644.265	1650.937	1657.612	1664.291	1670.972	1677.658	1684.346	1691.038	1697.734	1704.433	290

These tables of $f \ln f$ are useful in computing the log likelihood ratio G for multinomial distributions. G is used as a test statistic in tests of goodness of fit and independence in contingency tables (pp. 47, 48).

$f \ln f$ **(300–599)**

f	0	1	2	3	4	5	6	7	8	9	f
300	1711.135	1717.840	1724.549	1731.261	1737.976	1744.695	1751.417	1758.142	1764.871	1771.602	300
310	1778.337	1785.076	1791.817	1798.562	1805.309	1812.060	1818.815	1825.572	1832.332	1839.096	310
320	1845.863	1852.633	1859.406	1866.182	1872.961	1879.743	1886.529	1893.317	1900.108	1906.903	320
330	1913.701	1920.501	1927.305	1934.111	1940.921	1947.734	1954.549	1961.368	1968.190	1975.014	330
340	1981.842	1988.672	1995.505	2002.342	2009.181	2016.023	2022.868	2029.716	2036.566	2043.420	340
350	2050.277	2057.136	2063.998	2070.863	2077.731	2084.602	2091.475	2098.352	2105.231	2112.113	350
360	2118.997	2125.885	2132.775	2139.668	2146.564	2153.463	2160.364	2167.268	2174.175	2181.084	360
370	2187.996	2194.911	2201.829	2208.749	2215.672	2222.597	2229.526	2236.456	2243.390	2250.326	370
380	2257.265	2264.207	2271.151	2278.097	2285.047	2291.999	2298.953	2305.910	2312.870	2319.832	380
390	2326.797	2333.765	2340.735	2347.707	2354.682	2361.660	2368.640	2375.623	2382.608	2389.596	390
400	2396.586	2403.579	2410.574	2417.571	2424.572	2431.574	2438.579	2445.587	2452.597	2459.609	400
410	2466.624	2473.642	2480.662	2487.684	2494.709	2501.736	2508.765	2515.797	2522.831	2529.868	410
420	2536.907	2543.948	2550.992	2558.038	2565.087	2572.138	2579.191	2586.247	2593.305	2600.365	420
430	2607.428	2614.493	2621.560	2628.629	2635.701	2642.776	2649.852	2656.931	2664.012	2671.095	430
440	2678.181	2685.269	2692.359	2699.451	2706.546	2713.643	2720.742	2727.844	2734.947	2742.053	440
450	2749.161	2756.272	2763.384	2770.499	2777.616	2784.735	2791.857	2798.980	2806.106	2813.234	450
460	2820.364	2827.496	2834.631	2841.768	2848.906	2856.047	2863.191	2870.336	2877.483	2884.633	460
470	2891.784	2898.938	2906.094	2913.252	2920.412	2927.575	2934.739	2941.905	2949.074	2956.245	470
480	2963.417	2970.592	2977.769	2984.948	2992.129	2999.312	3006.497	3013.685	3020.874	3028.065	480
490	3035.259	3042.454	3049.652	3056.851	3064.053	3071.256	3078.462	3085.669	3092.879	3100.090	490
500	3107.304	3114.520	3121.737	3128.957	3136.178	3143.402	3150.628	3157.855	3165.085	3172.316	500
510	3179.549	3186.785	3194.022	3201.262	3208.503	3215.746	3222.991	3230.238	3237.487	3244.738	510
520	3251.991	3259.246	3266.502	3273.761	3281.022	3288.284	3295.548	3302.815	3310.083	3317.353	520
530	3324.625	3331.899	3339.174	3346.452	3353.731	3361.013	3368.296	3375.581	3382.868	3390.157	530
540	3397.447	3404.740	3412.034	3419.330	3426.628	3433.928	3441.230	3448.533	3455.839	3463.146	540
550	3470.455	3477.766	3485.079	3492.393	3499.709	3507.027	3514.347	3521.669	3528.992	3536.318	550
560	3543.645	3550.973	3558.304	3565.636	3572.971	3580.307	3587.644	3594.984	3602.325	3609.668	560
570	3617.013	3624.359	3631.708	3639.058	3646.409	3653.763	3661.118	3668.475	3675.834	3683.194	570
580	3690.556	3697.920	3705.286	3712.653	3720.022	3727.393	3734.765	3742.140	3749.515	3756.893	580
590	3764.272	3771.653	3779.036	3786.420	3793.806	3801.194	3808.583	3815.975	3823.367	3830.762	590

$f \ln f$ **(600–899)**

f	0	1	2	3	4	5	6	7	8	9	f
600	3838.158	3845.556	3852.955	3860.356	3867.759	3875.163	3882.569	3889.977	3897.386	3904.797	600
610	3912.210	3919.624	3927.040	3934.458	3941.877	3949.298	3956.720	3964.144	3971.570	3978.997	610
620	3986.426	3993.857	4001.289	4008.722	4016.158	4023.595	4031.033	4038.473	4045.915	4053.359	620
630	4060.803	4068.250	4075.698	4083.148	4090.599	4098.052	4105.506	4112.962	4120.420	4127.879	630
640	4135.340	4142.802	4150.266	4157.731	4165.198	4172.666	4180.136	4187.608	4195.081	4202.556	640
650	4210.032	4217.510	4224.989	4232.470	4239.952	4247.436	4254.921	4262.408	4269.897	4277.387	650
660	4284.878	4292.371	4299.866	4307.362	4314.859	4322.358	4329.859	4337.361	4344.864	4352.370	660
670	4359.876	4367.384	4374.894	4382.405	4389.917	4397.431	4404.947	4412.463	4419.982	4427.502	670
680	4435.023	4442.546	4450.070	4457.596	4465.123	4472.652	4480.182	4487.714	4495.247	4502.781	680
690	4510.317	4517.855	4525.393	4532.934	4540.476	4548.019	4555.563	4563.109	4570.657	4578.206	690
700	4585.756	4593.308	4600.861	4608.416	4615.972	4623.529	4631.088	4638.649	4646.210	4653.774	700
710	4661.338	4668.904	4676.471	4684.040	4691.610	4699.182	4706.755	4714.329	4721.905	4729.482	710
720	4737.061	4744.641	4752.222	4759.805	4767.389	4774.974	4782.561	4790.150	4797.739	4805.330	720
730	4812.923	4820.516	4828.111	4835.708	4843.306	4850.905	4858.505	4866.107	4873.711	4881.315	730
740	4888.921	4896.528	4904.137	4911.747	4919.359	4926.971	4934.585	4942.201	4949.817	4957.435	740
750	4965.055	4972.676	4980.298	4987.921	4995.546	5003.172	5010.799	5018.428	5026.058	5033.689	750
760	5041.322	5048.956	5056.591	5064.228	5071.866	5079.505	5087.146	5094.787	5102.431	5110.075	760
770	5117.721	5125.368	5133.016	5140.666	5148.317	5155.969	5163.622	5171.277	5178.933	5186.591	770
780	5194.249	5201.909	5209.570	5217.233	5224.897	5232.562	5240.228	5247.896	5255.564	5263.235	780
790	5270.906	5278.579	5286.253	5293.928	5301.604	5309.282	5316.961	5324.641	5332.323	5340.005	790
800	5347.689	5355.375	5363.061	5370.749	5378.438	5386.128	5393.819	5401.512	5409.206	5416.901	800
810	5424.598	5432.295	5439.994	5447.694	5455.396	5463.098	5470.802	5478.507	5486.213	5493.921	810
820	5501.630	5509.339	5517.051	5524.763	5532.477	5540.191	5547.907	5555.624	5563.343	5571.063	820
830	5578.783	5586.505	5594.229	5601.953	5609.679	5617.405	5625.134	5632.863	5640.593	5648.325	830
840	5656.058	5663.792	5671.527	5679.263	5687.001	5694.739	5702.479	5710.220	5717.963	5725.706	840
850	5733.451	5741.197	5748.944	5756.692	5764.441	5772.192	5779.943	5787.696	5795.450	5803.206	850
860	5810.962	5818.719	5826.478	5834.238	5841.999	5849.761	5857.524	5865.289	5873.054	5880.821	860
870	5888.589	5896.358	5904.128	5911.900	5919.672	5927.446	5935.221	5942.997	5950.774	5958.552	870
880	5966.331	5974.112	5981.893	5989.676	5997.460	6005.245	6013.031	6020.818	6028.607	6036.396	880
890	6044.187	6051.979	6059.772	6067.566	6075.361	6083.157	6090.955	6098.753	6106.553	6114.353	890

LOWER QUANTILES $T_{[P]}$ OF THE WILCOXON SIGNED-RANK TEST STATISTIC

n	P .005	.01	.025	.05	.10	n	P .005	.01	.025	.05	n	P .005	.01	.025	.05
5	0	0	0	1	3										
6	0	0	1	3	4	21	43	50	59	68	36	172	186	209	228
7	0	1	3	4	6	22	49	56	66	76	37	183	199	222	242
8	1	2	4	6	9	23	55	63	74	84	38	195	212	236	257
9	2	4	6	9	11	24	62	70	82	92	39	208	225	250	272
10	4	6	9	11	15	25	69	77	90	101	40	221	239	265	287
11	6	8	11	14	18	26	76	85	99	111	41	234	253	280	303
12	8	10	14	18	22	27	84	93	108	120	42	248	267	295	320
13	10	13	18	22	27	28	92	102	117	131	43	262	282	311	337
14	13	16	22	26	32	29	101	111	127	141	44	277	297	328	354
15	16	20	26	31	37	30	110	121	138	152	45	292	313	344	372
16	20	24	30	36	43	31	119	131	148	164	46	308	329	362	390
17	24	28	35	42	49	32	129	141	160	176	47	323	346	379	408
18	28	33	41	48	56	33	139	152	171	188	48	340	363	397	427
19	33	38	47	54	63	34	149	163	183	201	49	356	380	416	447
20	38	44	53	61	70	35	160	174	196	214	50	374	398	435	467

Source: R. L. McCornack, *Journal of the American Statistical Association* **60** (1965), 864–71, Table 1.

Upper quantiles are given by:

$$T_{[P]} = \tfrac{1}{2}n(n+1) - T_{[1-P]}$$

The mean and variance of the sampling distribution of T are:

$$\mu_T = n(n+1)/4, \quad \sigma_T^2 = n(n+1)(2n+1)/24$$

For large n the distribution of T is approximately normal. The distribution of $(T - \mu_T)/\sigma_T$ is approximately N(0, 1), and

$$T_{[P]} \approx \mu_T + \sigma_T z_{[P]}$$

where $z_{[P]}$ is a quantile of N(0, 1).

The Wilcoxon signed-rank test for matched pairs. Given n pairs (x_i, y_i), the differences $d_i = x_i - y_i$, which must be on at least an ordinal scale, are ranked in order of increasing absolute magnitude (i.e. without regard to sign) from 1 to n. To test the null hypothesis H_0 that the d-distribution is symmetric with median 0, against the alternative hypothesis H_1 that the median is *negative* (so that X-values tend to be smaller than Y-values) – a *one-sided test* – we use as test statistic T the sum R_+ of the ranks of the positive differences, and reject H_0 in favour of H_1 if T is less than the appropriate lower quantile.

If H_1 is the hypothesis that the median of the d-distribution is *positive* we take T to be R_-, the sum of the ranks of the negative differences. Note that $R_+ + R_- = \tfrac{1}{2}n(n+1)$.

If H_1 is the hypothesis that the d-distribution has a *non-zero* median – the *two-sided test* – we use as test statistic T^*, the smaller of R_+ and R_-. The quantiles of T^* are related to those of T. For $P \leqslant \tfrac{1}{2}$:

$$T_{[P]} = T^*_{[2P]}$$

The test is sensitive to differences of location, and rejection of the hypothesis usually implies inequality of the medians M_X and M_Y. See Example 2 on p. 66.

The Wilcoxon signed-rank median test. To test the hypothesis that a sample $(x_1, x_2, \ldots, x_n)$ has been drawn from a population symmetric about the median M, put $x_i - M = d_i$ and proceed as above.

LOWER QUANTILES OF THE MANN–WHITNEY TEST STATISTIC U

m	n	.001	.005	.01	.025	.05	.10		m	n	.001	.005	.01	.025	.05	.10
2	2	0	0	0	0	0	0		5	7	0	2	4	6	7	9
	3	0	0	0	0	0	1			8	1	3	5	7	9	11
	4	0	0	0	0	0	1			9	2	4	6	8	10	13
	5	0	0	0	0	1	2			10	2	5	7	9	12	14
	6	0	0	0	0	1	2			11	3	6	8	10	13	16
	7	0	0	0	0	1	2			12	3	7	9	12	14	18
	8	0	0	0	1	2	3			13	4	8	10	13	16	19
	9	0	0	0	1	2	3			14	4	8	11	14	17	21
	10	0	0	0	1	2	4			15	5	9	12	15	19	23
	11	0	0	0	1	2	4			16	6	10	13	16	20	24
	12	0	0	0	2	3	5			17	6	11	14	18	21	26
	13	0	0	1	2	3	5			18	7	12	15	19	23	28
	14	0	0	1	2	4	6			19	8	13	16	20	24	29
	15	0	0	1	2	4	6			20	8	14	17	21	26	31
	16	0	0	1	2	4	6		6	6	0	3	4	6	8	10
	17	0	0	1	3	4	7			7	1	4	5	7	9	12
	18	0	0	1	3	5	7			8	2	5	7	9	11	14
	19	0	1	2	3	5	8			9	3	6	8	11	13	16
	20	0	1	2	3	5	8			10	4	7	9	12	15	18
3	3	0	0	0	0	1	2			11	5	8	10	14	17	20
	4	0	0	0	0	1	2			12	5	10	12	15	18	22
	5	0	0	0	1	2	3			13	6	11	13	17	20	24
	6	0	0	0	2	3	4			14	7	12	14	18	22	26
	7	0	0	1	2	3	5			15	8	13	16	20	24	28
	8	0	0	1	3	4	6			16	9	14	17	22	26	30
	9	0	1	2	3	5	6			17	10	16	19	23	27	32
	10	0	1	2	4	5	7			18	11	17	20	25	29	35
	11	0	1	2	4	6	8			19	12	18	21	26	31	37
	12	0	2	3	5	6	9			20	13	19	23	28	33	39
	13	0	2	3	5	7	10		7	7	2	5	7	9	12	14
	14	0	2	3	6	8	11			8	3	7	8	11	14	17
	15	0	3	4	6	8	11			9	4	8	10	13	16	19
	16	0	3	4	7	9	12			10	6	10	12	15	18	22
	17	1	3	5	7	10	13			11	7	11	13	17	20	24
	18	1	3	5	8	10	14			12	8	13	15	19	22	27
	19	1	4	5	8	11	15			13	9	14	17	21	25	29
	20	1	4	6	9	12	16			14	10	16	18	23	27	32
4	4	0	0	0	1	2	4			15	11	17	20	25	29	34
	5	0	0	1	2	3	5			16	12	19	22	27	31	37
	6	0	1	2	3	4	6			17	14	20	24	29	34	39
	7	0	1	2	4	5	7			18	15	22	25	31	36	42
	8	0	2	3	5	6	8			19	16	23	27	33	38	44
	9	0	2	4	5	7	10			20	17	25	29	35	40	47
	10	1	3	4	6	8	11		8	8	5	8	10	14	16	20
	11	1	3	5	7	9	12			9	6	10	12	16	19	23
	12	1	4	6	8	10	13			10	7	12	14	18	21	25
	13	2	4	6	9	11	14			11	9	14	16	20	24	28
	14	2	5	7	10	12	16			12	10	16	18	23	27	31
	15	2	6	8	11	13	17			13	12	18	21	25	29	34
	16	3	6	8	12	15	18			14	13	19	23	27	32	37
	17	3	7	9	12	16	19			15	15	21	25	30	34	40
	18	4	7	10	13	17	21			16	16	23	27	32	37	43
	19	4	8	10	14	18	22			17	18	25	29	35	40	46
	20	4	9	11	15	19	23			18	19	27	31	37	42	49
5	5	0	1	2	3	5	6			19	21	29	33	39	45	52
	6	0	2	3	4	6	8			20	22	31	35	42	48	55

LOWER QUANTILES OF THE MANN–WHITNEY TEST STATISTIC U

m	n	P .001	.005	.01	.025	.05	.10	m	n	P .001	.005	.01	.025	.05	.10
9	9	8	12	15	18	22	26	12	18	38	48	54	62	69	78
	10	9	14	17	21	25	29		19	41	52	57	66	73	82
	11	11	17	19	24	28	32		20	43	55	61	70	78	87
	12	13	19	22	27	31	36	13	13	27	35	40	46	52	59
	13	15	21	24	29	34	39		14	30	39	44	51	57	64
	14	16	23	27	32	37	42		15	33	43	48	55	62	69
	15	18	25	29	35	40	46		16	36	46	52	60	66	75
	16	20	28	32	38	43	49		17	39	50	56	64	71	80
	17	22	30	34	40	46	53		18	43	54	60	68	76	85
	18	24	32	37	43	49	56		19	46	58	64	73	81	90
	19	26	34	39	46	52	59		20	49	61	68	77	85	95
	20	27	37	41	49	55	63	14	14	33	43	48	56	62	70
10	10	11	17	20	24	28	33		15	37	47	52	60	67	75
	11	13	19	23	27	32	37		16	40	51	57	65	72	81
	12	15	22	25	30	35	40		17	44	55	61	70	78	86
	13	18	25	28	34	38	44		18	47	59	66	75	83	92
	14	20	27	31	37	42	48		19	51	64	70	79	88	98
	15	22	30	34	40	45	52		20	55	68	74	84	93	103
	16	24	32	37	43	49	55	15	15	41	52	57	65	73	81
	17	26	35	39	46	52	59		16	44	56	62	71	78	87
	18	28	38	42	49	56	63		17	48	61	67	76	84	93
	19	30	40	45	53	59	67		18	52	65	71	81	89	99
	20	33	43	48	56	63	71		19	56	70	76	86	95	105
11	11	16	22	26	31	35	41		20	60	74	81	91	101	111
	12	18	25	29	34	39	45	16	16	49	61	67	76	84	94
	13	21	28	32	38	43	49		17	53	66	72	82	90	100
	14	23	31	35	41	47	53		18	57	71	77	87	96	107
	15	25	34	38	45	51	58		19	61	75	83	93	102	113
	16	28	37	42	48	55	62		20	66	80	88	99	108	120
	17	30	40	45	52	58	66	17	17	58	71	78	88	97	107
	18	33	43	48	56	62	70		18	62	76	83	94	103	114
	19	35	46	51	59	66	74		19	67	82	89	100	110	121
	20	38	49	54	63	70	79		20	71	87	94	106	116	128
12	12	21	28	32	38	43	50	18	18	67	82	89	100	110	121
	13	24	32	36	42	48	54		19	72	88	95	107	117	129
	14	26	35	39	46	52	59		20	77	93	101	113	124	136
	15	29	38	43	50	56	64	19	19	78	94	102	114	124	136
	16	32	42	47	54	61	68		20	83	100	108	120	131	144
	17	35	45	50	58	65	73	20	20	89	106	115	128	139	152

Source: For the tables on pp. 64, 65, L. R. Verdooren, *Biometrika* **50** (1963), 177–86. This is reproduced in *Biometrika Tables for Statisticians*, vol. 2, Table 22.

The table gives lower quantiles $U_{[P]}$ of the null distribution of the test statistic U. Note that m and n are interchangeable. Upper quantiles are given by:

$$U_{[P]} = mn - U_{[1-P]}$$

The mean and variance of the distribution are:

$$\mu_U = mn/2 \quad \sigma_U^2 = mn(m+n+1)/12$$

The distribution of U is approximately normal if m or $n > 20$. The approximation

$$U_{[P]} \approx mn/2 + \sqrt{\{mn(m+n+1)(m+n-1)/12\}}\, r_{[P]}$$

(where r is the product-moment correlation coefficient with $\nu = m+n-2$ degrees of freedom) is somewhat closer, but less convenient. For a description of the U-test see p. 66.

The Wilcoxon/Mann–Whitney rank-sum test. This is a test for differences of location between two populations. Independent random samples $x_1, x_2, \ldots, x_m$ and $y_1, y_2, \ldots, y_n$ are given. To test the null hypothesis H_0 that $\mathrm{Prob}(x > y) = \mathrm{Prob}(x < y)$ against the alternative *one-sided* hypothesis that $\mathrm{Prob}(x > y) < \mathrm{Prob}(x < y)$ (so that the xs tend to be smaller than the ys) we use as test statistic U, the number of pairs x_i, y_j for which $x_i > y_j$. If U is less than the appropriate lower quantile we reject H_0 and accept H_1.

Similarly if H_1 is the hypothesis that $\mathrm{Prob}(x > y) > \mathrm{Prob}(x < y)$ we use as test statistic U', the number of pairs for which $x_i < y_j$. U' and U have the same null distribution, so that $U'_{[P]} = U_{[P]}$. Note that (ignoring ties) $U + U' = mn$.

Equivalently, if the xs and ys are pooled and ranked in ascending order of magnitude from 1 to $m+n$, we define

$$U = R_X - \tfrac{1}{2}m(m+1) \quad U' = R_Y - \tfrac{1}{2}n(n+1)$$

where R_X and R_Y are the sums of the ranks of the xs and ys respectively.

For the *two-sided* test with $H_1: \mathrm{Prob}(x > y) \neq \mathrm{Prob}(x < y)$ we use as test statistic U^*, the smaller of U and U'. The quantiles of U^* are related to those of U. For $P \leqslant \tfrac{1}{2}$:

$$U_{[P]} = U^*_{[2P]} \quad (\text{e.g. } U^*_{[.05]} = U_{[.025]})$$

The test was introduced by Wilcoxon, who used R_Y as test statistic.

Example 1. The performance of an electric switch can be measured by the number of times it can be operated before it fails. In order to decide whether the performance of switches of type A is significantly better than that of cheaper switches of type B, the performances of 6 switches of type A and 8 of type B were compared by repeatedly operating them and noting the order in which they failed. The failure order constitutes a performance ranking. The observed order is set out below.

Type	B	B	B	B	B	A	A	B	A	A	A	B	B	A
Failure order	1	2	3	4	5	6	7	8	9	10	11	12	13	14

A one-sided Mann–Whitney test can be used, with $m = 6$ (type A) and $n = 8$ (type B). The appropriate test statistic is U':

$$U' = R_B - \tfrac{1}{2}n(n+1) = 48 - 36 = 12$$

(R_B = sum of ranks of switches of type B). Now, for $m = 6$ and $n = 8$

$$U'_{[.05]} = U_{[.05]} = 11 \quad U'_{[.10]} = U_{[.10]} = 14$$

We conclude that the performance of switches of type A is significantly better at level 0.1 but not at 0.05.

Alternatively the one-sided Smirnov test (p. 69) or the two-sample runs test (p. 77) can be used.

Example 2. It is desired to test at 10% significance level whether treatment with a drug affects the mean reaction times. Eight individuals ($i = 1, 2, \ldots, 8$) are selected randomly; x and y are the reaction times before and after treatment.

i	1	2	3	4	5	6	7	8	
x_i	81	79	93	71	86	82	82	75	$R_+ = 30$
y_i	72	76	78	73	75	86	77	74	$R_- = 6$
$x_i - y_i$	+9	+3	+15	−2	+11	−4	+5	+1	$T^* = \min(R_+, R_-)$
R_i	6	3	8	2	7	4	5	1	$= 6$

The X- and Y-distributions will be identical and the d-distribution symmetric if the drug is ineffective; the Wilcoxon signed-rank test may therefore be used. We require a two-sided test, and use the test statistic T^* which has the value 6. Since, for $n = 8$, $T^*_{[.10]} = T_{[.05]} = 6$ we conclude that the change in mean reaction times is not significant at level 0.1.

UPPER QUANTILES FOR KOLMOGOROV TESTS

	P for one-sided tests						P for one-sided tests			
	.90	.95	.975	.99	.995		.90	.95	.975	.99	.995
		P for two-sided tests						P for two-sided tests			
n		.90	.95	.98	.99	n		.90	.95	.98	.99
1	.900	.950	.975	.990	.995	31	.187	.214	.238	.266	.285
2	.684	.776	.842	.900	.929	32	.184	.211	.234	.262	.281
3	.565	.636	.708	.785	.829	33	.182	.208	.231	.258	.277
4	.493	.565	.624	.689	.734	34	.179	.205	.227	.254	.273
5	.447	.509	.563	.627	.669	35	.177	.202	.224	.251	.269
6	.410	.468	.519	.577	.617	36	.174	.199	.221	.247	.265
7	.381	.436	.483	.538	.576	37	.172	.196	.218	.244	.262
8	.358	.410	.454	.507	.542	38	.170	.194	.215	.241	.258
9	.339	.387	.430	.480	.513	39	.168	.191	.213	.238	.255
10	.323	.369	.409	.457	.489	40	.165	.189	.210	.235	.252
11	.308	.352	.391	.437	.468	41	.163	.187	.208	.232	.249
12	.296	.338	.375	.419	.449	42	.162	.185	.205	.229	.246
13	.285	.325	.361	.404	.432	43	.160	.183	.203	.227	.243
14	.275	.314	.349	.390	.418	44	.158	.181	.201	.224	.241
15	.266	.304	.338	.377	.404	45	.156	.179	.198	.222	.238
16	.258	.295	.327	.366	.392	46	.155	.177	.196	.219	.235
17	.250	.286	.318	.355	.381	47	.153	.175	.194	.217	.233
18	.244	.279	.309	.346	.371	48	.151	.173	.192	.215	.231
19	.237	.271	.301	.337	.361	49	.150	.171	.190	.213	.228
20	.232	.265	.294	.329	.352	50	.148	.170	.188	.211	.226
21	.226	.259	.287	.321	.344	55	.142	.162	.180	.201	.216
22	.221	.253	.281	.314	.337	60	.136	.155	.172	.193	.207
23	.216	.247	.275	.307	.330	65	.131	.149	.166	.185	.199
24	.212	.242	.269	.301	.323	70	.126	.144	.160	.179	.192
25	.208	.238	.264	.295	.317	75	.122	.139	.154	.173	.185
26	.204	.233	.259	.290	.311	80	.118	.135	.150	.167	.179
27	.200	.229	.254	.284	.305	85	.114	.131	.145	.162	.174
28	.197	.225	.250	.279	.300	90	.111	.127	.141	.158	.169
29	.193	.221	.246	.275	.295	95	.108	.124	.137	.154	.165
30	.190	.218	.242	.270	.290	100	.106	.121	.134	.150	.161
						k	1.073	1.224	1.358	1.517	1.628

Source: L. H. Miller, *Journal of the American Statistical Association* **51** (1956), 111–21, Table 1.

Kolmogorov tests are tests of goodness of fit. Let $S(x)$ denote the cumulative frequency distribution function for a given random sample $x_1, x_2, \ldots, x_n$ taken from a population with unknown cumulative probability distribution function $F(x)$. ($S(x)$ is defined as the number of $x_i \leqslant x$, divided by n). To test the null hypothesis $H_0 : F(x) = F_0(x)$ for all x, where $F_0(x)$ is some hypothesised cumulative distribution function, against the *one-sided* alternative hypothesis $H_1 : F(x) < F_0(x)$ for some x, we use the test statistic

$$D^+ = \max\{F_0(x) - S(x)\}$$

and reject H_0 in favour of H_1 if D^+ exceeds the appropriate upper quantile.

If the alternative hypothesis is that $F(x) > F_0(x)$ for some x, we use the statistic

$$D^- = \max\{S(x) - F_0(x)\}$$

To test H_0 against the *two-sided* alternative $F(x) \neq F_0(x)$ for some x, we use the statistic

$$D = \max|S(x) - F_0(x)|$$

The table gives upper quantiles of D^+ and D^- for one-sided tests and D for two-sided tests. For large n, quantiles are given approximately by $k/\sqrt{n}$; values of k are given in the table above. For example, for $n = 160$, $D^+_{[.99]} = 1.517/\sqrt{160} = 0.120$; $D_{[.99]} = 1.628/\sqrt{160} = 0.129$.

(A) UPPER QUANTILES OF mnD FOR SMIRNOV TESTS ($m \neq n$)

n \ m	2	3	4	5	6	7	8	9	10	11	12	13	14	15	16	17	18	19	20	21	22	23	24	25	n
2		6	8	10	12	14	14	16	18	20	22	24	26	26	28	30	32	34	36	36	38	40	42	44	2
3			12	12	15	18	18	21	24	27	27	30	33	33	36	39	42	42	45	48	48	51	54	57	3
4				16	18	21	24	27	28	32	32	36	40	41	44	47	48	52	56	56	60	61	64	67	4
5		15	20		20	25	27	31	35	35	40	42	45	50	50	53	57	60	60	65	68	70	75	75	5
6		18	20	25		29	32	36	38	42	42	48	52	54	58	61	66	66	70	72	76	79	84	84	6
7		21	24	30	35		35	40	43	45	51	52	56	61	63	67	70	74	78	84	83	87	91	94	7
8	16	24	28	32	36	42		45	46	50	56	59	62	66	72	72	78	80	84	88	92	97	96	103	8
9	18	24	32	36	42	47	54		52	57	60	64	67	72	76	81	81	88	91	96	100	103	108	112	9
10	20	27	32	40	44	50	56	62		59	64	68	72	75	82	86	90	93	100	100	106	111	116	120	10
11	22	30	36	44	49	56	61	68	70		66	73	79	83	88	92	96	100	105	111	110	118	122	128	11
12	24	33	40	48	54	58	64	72	78	85		80	84	90	92	98	102	106	112	117	122	124	132	132	12
13	26	36	44	50	59	64	70	77	81	88	94		88	94	99	104	108	113	117	124	129	133	138	144	13
14	28	39	44	55	60	70	74	81	88	93	102	103		97	104	109	114	120	124	133	136	141	144	149	14
15	30	39	48	55	66	70	80	87	95	99	105	113	122		113	114	120	126	130	135	143	148	153	155	15
16	32	42	52	60	68	75	80	92	98	105	112	118	124	132		123	126	132	136	144	148	155	160	165	16
17	34	45	56	65	72	81	87	93	103	109	117	126	131	140	142		132	139	144	150	155	162	166	172	17
18	36	48	56	67	78	84	92	99	106	115	120	130	138	144	152	163		141	150	156	162	168	174	178	18
19	36	51	60	70	78	88	96	106	112	121	128	137	144	150	158	164	175		152	161	168	174	180	186	19
20	38	54	64	75	84	92	100	108	120	125	136	142	150	155	164	172	180	186		172	174	182	188	195	20
21	40	54	68	79	87	98	105	114	120	133	138	148	154	165	172	179	186	197	198		182	188	195	200	21
22	42	57	68	80	90	98	110	118	128	132	146	155	162	168	178	186	194	203	210	222		193	202	208	22
23	44	60	72	85	96	105	114	125	134	141	148	159	169	178	185	192	201	208	217	225	236		204	214	23
24	46	63	76	86	96	109	120	129	138	148	156	164	174	183	192	200	210	217	224	234	240	248		216	24
25	48	66	80	90	102	112	120	132	145	153	164	171	180	190	197	206	214	223	230	241	249	260	261		25
	2	3	4	5	6	7	8	9	10	11	12	13	14	15	16	17	18	19	20	21	22	23	24	25	

Source: *Selected Tables in Mathematical Statistics* (Institute of Mathematical Statistics), vol. 1, pp. 79–170.

The table gives upper quantiles $mnD_{[.95]}$ and $mnD_{[.99]}$ for the statistic D used in two-sided Smirnov tests. Quantiles $mnD_{[.95]}$ are placed above the diagonal line, quantiles $mnD_{[.99]}$ below it. Note that m and n are interchangeable (i.e. $D(m,n) = D(n,m)$). With adequate accuracy $D_{[.95]} = D^{+}_{[.975]}$ and $D_{[.99]} = D^{+}_{[.995]}$.

(B) UPPER QUANTILES OF nD AND nD^+ FOR SMIRNOV TESTS ($m = n$)

	P for one-sided tests						P for one-sided tests				
	.90	.95	.975	.99	.995		.90	.95	.975	.99	.995
		P for two-sided tests						P for two-sided tests			
n		.90	.95	.98	.99	n		.90	.95	.98	.99
						21	6	7	8	9	10
						22	7	8	8	10	10
3	2	3	3			23	7	8	9	10	10
4	3	3	3	4		24	7	8	9	10	11
5	3	3	4	4	4	25	7	8	9	10	11
6	3	4	4	5	5	26	7	8	9	10	11
7	4	4	5	5	5	27	7	8	9	11	11
8	4	4	5	5	6	28	8	9	10	11	12
9	4	5	5	6	6	29	8	9	10	11	12
10	4	5	6	6	7	30	8	9	10	11	12
11	5	5	6	7	7	31	8	9	10	11	12
12	5	5	6	7	7	32	8	9	10	12	12
13	5	6	6	7	8	34	8	10	11	12	13
14	5	6	7	7	8	36	9	10	11	12	13
15	5	6	7	8	8	38	9	10	11	13	14
16	6	6	7	8	9	40	9	10	12	13	14
17	6	7	7	8	9						
18	6	7	8	9	9	k	1.073	1.224	1.358	1.517	1.628
19	6	7	8	9	9						
20	6	7	8	9	10						

Source: Z. W. Birnbaum & R. A. Hall, *Annals of Mathematical Statistics* 31 (1960), 710–20.

These are tests of identity of populations. Let $S(x)$ denote the cumulative frequency distribution function for a given random sample $x_1, x_2, \ldots, x_m$ taken from a population with unknown cumulative probability distribution function $F(x)$, and let $S'(x)$ denote the cumulative frequency distribution function for an independent random sample $y_1, y_2, \ldots, y_n$ taken from a second population with unknown distribution function $F'(x)$. ($S(x)$ is defined as the number of $x_i \leqslant x$, divided by m; $S'(x)$ is similarly defined.) To test the null hypothesis $H_0 : F(x) = F'(x)$ for all x against the *one-sided* alternative $H_1 : F(x) < F'(x)$ for some x, we use as test statistic

$$D^+ = \max \{ S'(x) - S(x) \}$$

and reject H_0 in favour of H_1 if D^+ exceeds the appropriate quantile.

To test the null hypothesis $H_0 : F(x) = F'(x)$ for all x against the *two-sided* alternative $H_1 : F(x) \neq F'(x)$ for some x, we use as test statistic:

$$D = \max |S(x) - S'(x)|$$

Table (A) gives upper quantiles for unequal samples ($m \neq n$). Table (B) gives quantiles $nD_{[P]}$ and $nD^+_{[P]}$ for equal samples ($m = n$). For example, for $m = n = 20$, $20 D^+_{[.95]} = 7$, so that $D^+_{[.95]} = 7/20$ and therefore $\mathrm{Prob}(D^+ > 7/20) < 0.05$; similarly $20 D_{[.95]} = 8$ so that $D_{[.95]} = 8/20$ and $\mathrm{Prob}(D > 8/20) < 0.05$.

For large m and n quantiles of D and D^+ are approximately:

$$k \sqrt{\left(\frac{m+n}{mn} \right)}$$

Values of k are given in table (B). For example, for $m = 40$ and $n = 60$:

$$D_{[.99]} = 1.628 \sqrt{(100/2400)} = 0.332$$

69

n_1	n_2	n_3	H	Q	n_1	n_2	n_3	H	Q	n_1	n_2	n_3	H	Q
2	1	1	2.700	.500	4	3	2	4.444	.102	5	2	2	4.293	.122
2	2	1	3.600	.200				4.511	.098				4.373	.090
2	2	2	3.714	.200				5.400	.051				5.040	.056
			4.571	.067				5.444	.046				5.160	.034
3	1	1	3.200	.300				6.300	.011				6.133	.013
3	2	1	3.857	.133				6.444	.008				6.533	.008
			4.286	.100	4	3	3	4.700	.101	5	3	1	3.840	.123
3	2	2	4.464	.105				4.709	.092				4.018	.095
			4.500	.067				5.727	.050				4.871	.052
			4.714	.048				5.791	.046				4.960	.048
			5.357	.029				6.709	.013				6.400	.012
3	3	1	4.000	.129				6.746	.010	5	3	2	4.494	.101
			4.571	.100	4	4	1	4.067	.102				4.651	.091
			5.143	.043				4.167	.082				5.106	.052
3	3	2	4.250	.121				4.867	.054				5.251	.049
			4.556	.100				4.967	.048				6.822	.010
			5.139	.061				6.167	.022				6.909	.009
			5.361	.032				6.667	.010	5	3	3	4.412	.109
			6.250	.011	4	4	2	4.446	.103				4.533	.097
3	3	3	4.622	.100				4.554	.098				5.515	.051
			5.067	.086				5.236	.052				5.648	.049
			5.600	.050				5.454	.046				6.982	.011
			5.689	.029				6.873	.011				7.079	.009
			6.489	.011				7.036	.006	5	4	1	3.960	.102
			7.200	.004	4	4	3	4.477	.102				3.987	.098
4	1	1	3.571	.200				4.546	.099				4.860	.056
4	2	1	4.018	.114				5.576	.051				4.986	.044
			4.500	.076				5.598	.049				6.840	.011
			4.821	.057				7.136	.011				6.954	.008
4	2	2	4.167	.105				7.144	.010	5	4	2	4.518	.101
			4.458	.100	4	4	4	4.500	.104				4.541	.098
			5.125	.052				4.654	.097				5.268	.051
			5.333	.033				5.654	.054				5.273	.049
			6.000	.014				5.692	.049				7.118	.010
4	3	1	3.889	.129				7.538	.011				7.204	.009
			4.056	.093				7.654	.008	5	4	3	4.523	.103
			5.000	.057	5	1	1	3.857	.143				4.549	.099
			5.208	.050	5	2	1	4.050	.119				5.631	.050
			5.833	.021				4.200	.095				5.656	.049
								4.450	.071				7.395	.011
								5.000	.048				7.445	.010
								5.250	.036					

UPPER TAIL PROBABILITIES OF THE KRUSKAL–WALLIS TEST STATISTIC H

n_1	n_2	n_3	H	Q	n_1	n_2	n_3	H	Q	n_1	n_2	n_3	H	Q
5	4	4	4.553	.102	5	5	2	4.508	.100	5	5	4	4.520	.101
			4.619	.100				4.623	.097				4.523	.099
			5.618	.050				5.246	.051				5.643	.050
			5.657	.049				5.338	.047				5.666	.049
			7.744	.011				7.269	.010				7.791	.010
			7.760	.009				7.338	.010				7.823	.010
5	5	1	4.036	.105	5	5	3	4.536	.102	5	5	5	4.500	.102
			4.109	.086				4.545	.100				4.560	.100
			4.909	.053				5.626	.051				5.660	.051
			5.127	.046				5.706	.046				5.780	.049
			6.836	.011				7.543	.010				7.980	.010
			7.309	.009				7.578	.010				8.000	.009

Source: For the tables on pp. 70, 71, W. H. Kruskal & W. A. Wallis, *Journal of the American Statistical Association* 47 (1952), 585–621, Table 6.1 and 48 (1953), 907–11.

The Kruskal–Wallis test. This is a test of the hypothesis that k given independent samples are random samples from identical populations. Let n_j be the number of observations in sample j. The $N = \Sigma n_j$ observations (which must be on at least an ordinal scale) are pooled and ranked 1 to N. Let R_j be the sum of the ranks of the observations in sample j. The test statistic is:

$$H = \frac{12}{N(N+1)} \Sigma_j \frac{R_j^2}{n_j} - 3(N+1)$$

We reject the hypothesis if H exceeds the appropriate upper quantile. The test is sensitive to differences of location.

The table gives selected upper tail probabilities Q (near 0.10, 0.05 and 0.01) for the null distribution of H for $k = 3$ and $n \leqslant 5$. For larger sample sizes the distributions approximate to the chi-square distributions $\chi^2(k-1)$.

For $k = 2$ the test reduces to the Wilcoxon/Mann–Whitney two-sided rank-sum test.

UPPER TAIL PROBABILITIES Q FOR THE TERRY–HOEFFDING NORMAL SCORES TEST STATISTIC T

N	m	.001	.005	.01	.025	.05	.10
6	3					2.111 (.050)	1.707 (.100)
7	2					2.110 (.048)	1.705 (.095)
7	3				2.462 (.029)	2.110 (.057)	1.705 (.114)
8	2				2.276 (.036)	1.896 (.071)	1.576 (.107)
8	3			2.749 (.018)	2.428 (.036)	2.123 (.054)	1.744 (.107)
8	4			2.901 (.014)	2.596 (.029)	2.276 (.057)	1.896 (.100)
9	2				2.417 (.028)	2.057 (.056)	1.504 (.111)
9	3			2.989 (.012)	2.692 (.024)	2.332 (.048)	1.782 (.095)
9	4		3.264 (.0079)	2.989 (.016)	2.715 (.024)	2.417 (.048)	1.782 (.103)
10	2				2.540 (.022)	2.195 (.044)	1.657 (.111)
10	3		3.196 (.0083)	2.916 (.017)	2.663 (.025)	2.317 (.050)	1.819 (.100)
10	4		3.572 (.0048)	3.319 (.010)	2.820 (.024)	2.448 (.052)	1.942 (.095)
10	5		3.695 (.0040)	3.449 (.008)	2.916 (.028)	2.571 (.048)	2.033 (.103)
11	2			2.648 (.018)	2.315 (.036)	2.048 (.055)	1.791 (.091)
11	3		3.377 (.0061)	3.110 (.012)	2.777 (.024)	2.315 (.048)	1.853 (.097)
11	4	3.839 (.0030)	3.602 (.0061)	3.377 (.009)	2.915 (.024)	2.540 (.048)	2.028 (.100)
11	5	4.064 (.0022)	3.839 (.0043)	3.602 (.009)	3.002 (.026)	2.606 (.050)	2.078 (.097)
12	2			2.745 (.015)	2.422 (.030)	2.166 (.045)	1.653 (.106)
12	3		3.538 (.0046)	3.282 (.009)	2.848 (.023)	2.433 (.050)	1.885 (.100)
12	4	4.075 (.0020)	3.850 (.0040)	3.435 (.010)	3.001 (.024)	2.548 (.051)	2.055 (.099)
12	5	4.387 (.0013)	3.953 (.0051)	3.697 (.009)	3.104 (.025)	2.655 (.051)	2.155 (.098)
12	6	4.489 (.0011)	3.865 (.0054)	3.640 (.011)	3.179 (.026)	2.745 (.052)	2.166 (.102)
13	2			2.832 (.013)	2.518 (.026)	2.056 (.051)	1.668 (.103)
13	3	3.682 (.0035)	3.435 (.0070)	3.220 (.010)	2.832 (.024)	2.444 (.049)	1.883 (.101)
13	4	4.285 (.0014)	3.823 (.0056)	3.509 (.010)	3.047 (.025)	2.618 (.050)	2.073 (.098)
13	5	4.673 (.0008)	4.070 (.0047)	3.700 (.009)	3.237 (.024)	2.776 (.050)	2.204 (.100)
13	6	4.673 (.0012)	4.087 (.0052)	3.823 (.010)	3.294 (.025)	2.832 (.051)	2.247 (.100)
14	2			2.911 (.011)	2.605 (.022)	2.109 (.055)	1.663 (.099)
14	3	3.812 (.0028)	3.573 (.0055)	3.266 (.011)	2.823 (.025)	2.453 (.049)	1.943 (.099)
14	4	4.474 (.0010)	3.901 (.0050)	3.634 (.010)	3.117 (.025)	2.672 (.050)	2.159 (.098)
14	5	4.471 (.0010)	4.168 (.0050)	3.812 (.010)	3.269 (.025)	2.823 (.050)	2.250 (.099)
14	6	4.842 (.0010)	4.286 (.0047)	3.913 (.010)	3.395 (.025)	2.911 (.052)	2.315 (.100)
14	7	4.930 (.0012)	4.356 (.0049)	3.992 (.010)	3.445 (.025)	2.948 (.050)	2.337 (.100)
15	2			2.984 (.010)	2.451 (.029)	2.196 (.048)	1.736 (.095)
15	3	3.932 (.0022)	3.699 (.0044)	3.319 (.011)	2.910 (.024)	2.468 (.051)	1.963 (.099)
15	4	4.646 (.0007)	3.932 (.0051)	3.665 (.010)	3.183 (.025)	2.696 (.050)	2.151 (.100)
15	5	4.812 (.0010)	4.249 (.0050)	3.914 (.010)	3.382 (.025)	2.881 (.050)	2.285 (.101)
15	6	4.997 (.0010)	4.432 (.0050)	4.068 (.010)	3.518 (.025)	2.985 (.050)	2.366 (.100)
15	7	5.162 (.0009)	4.476 (.0050)	4.112 (.010)	3.567 (.025)	3.047 (.050)	2.422 (.100)
16	2		3.051 (.0083)	2.756 (.017)	2.529 (.025)	2.162 (.050)	1.689 (.100)
16	3	4.041 (.0018)	3.621 (.0054)	3.326 (.011)	2.925 (.025)	2.481 (.050)	1.959 (.100)
16	4	4.611 (.0011)	4.048 (.0049)	3.722 (.010)	3.230 (.025)	2.756 (.050)	2.174 (.100)
16	5	5.007 (.0009)	4.323 (.0050)	3.988 (.010)	3.458 (.024)	2.937 (.050)	2.318 (.100)
16	6	5.140 (.0010)	4.521 (.0050)	4.157 (.010)	3.591 (.025)	3.067 (.050)	2.421 (.100)
16	7	5.357 (.0010)	4.642 (.0050)	4.251 (.010)	3.677 (.025)	3.152 (.050)	2.481 (.100)
16	8	5.434 (.0009)	4.667 (.0050)	4.292 (.010)	3.722 (.025)	3.160 (.050)	2.512 (.100)
17	2		3.113 (.0074)	2.823 (.015)	2.601 (.022)	2.126 (.051)	1.649 (.103)
17	3	4.142 (.0015)	3.732 (.0044)	3.408 (.010)	2.967 (.025)	2.494 (.050)	1.980 (.100)
17	4	4.762 (.0008)	4.082 (.0050)	3.774 (.010)	3.265 (.025)	2.788 (.050)	2.199 (.100)
17	5	5.057 (.0010)	4.394 (.0050)	4.066 (.010)	3.481 (.025)	2.969 (.050)	2.353 (.100)

Source: For the tables on pp. 72, 73, J. H. Klotz, *Journal of the American Statistical Association* **59** (1964), 652–64.

UPPER TAIL PROBABILITIES Q FOR THE TERRY–HOEFFDING NORMAL SCORES TEST STATISTIC T

N	m	Q (nominal) .001	.005	.01	.025	.05	.10
17	6	5.286	4.612	4.244	3.659	3.119	2.458
		(.0010)	(.0050)	(.010)	(.025)	(.050)	(.100)
	7	5.432	4.762	4.377	3.774	3.220	2.540
		(.0010)	(.0051)	(.010)	(.025)	(.050)	(.100)
	8	5.559	4.835	4.444	3.818	3.265	2.578
		(.0010)	(.0050)	(.010)	(.025)	(.050)	(.100)
18	2		3.170	2.886	2.485	2.171	1.731
			(.0065)	(.013)	(.026)	(.052)	(.098)
	3	4.236	3.734	3.387	2.963	2.529	1.983
		(.0012)	(.0049)	(.010)	(.025)	(.050)	(.100)
	4	4.738	4.167	3.803	3.309	2.807	2.200
		(.0010)	(.0049)	(.010)	(.025)	(.050)	(.100)
	5	5.109	4.451	4.112	3.545	3.019	2.387
		(.0011)	(.0050)	(.010)	(.025)	(.050)	(.100)
	6	5.393	4.693	4.337	3.738	3.175	2.505
		(.0010)	(.0050)	(.010)	(.025)	(.050)	(.100)
	7	5.598	4.877	4.470	3.849	3.277	2.590
		(.0010)	(.0050)	(.010)	(.025)	(.050)	(.100)
	8	5.692	4.965	4.554	3.929	3.345	2.639
		(.0010)	(.0050)	(.010)	(.025)	(.050)	(.100)
	9	5.749	5.010	4.594	3.960	3.376	2.664
		(.0010)	(.0050)	(.010)	(.025)	(.050)	(.100)
19	2		3.224	2.944	2.551	2.108	1.714
			(.0058)	(.012)	(.023)	(.053)	(.099)
	3	4.324	3.772	3.437	2.972	2.533	2.002
		(.0010)	(.0052)	(.010)	(.025)	(.050)	(.100)
	4	4.817	4.195	3.847	3.338	2.822	2.249
		(.0010)	(.0049)	(.010)	(.025)	(.050)	(.099)

N	m	Q (nominal) .001	.005	.01	.025	.05	.10
19	5	5.210	4.512	4.174	3.583	3.062	2.401
		(.0010)	(.0050)	(.010)	(.025)	(.050)	(.100)
	6	5.495	4.791	4.387	3.781	3.226	2.542
		(.0010)	(.0050)	(.010)	(.025)	(.050)	(.100)
	7	5.757	4.963	4.568	3.929	3.350	2.642
		(.0010)	(.0050)	(.010)	(.025)	(.050)	(.100)
	8	5.888	5.086	4.669	4.026	3.420	2.693
		(.0010)	(.0050)	(.010)	(.025)	(.050)	(.100)
	9	5.929	5.164	4.739	4.065	3.459	2.731
		(.0010)	(.0050)	(.010)	(.025)	(.050)	(.100)
20	2		3.275	2.998	2.539	2.182	1.721
			(.0053)	(.011)	(.026)	(.047)	(.100)
	3	4.406	3.744	3.462	2.987	2.550	2.014
		(.0009)	(.0053)	(.010)	(.025)	(.050)	(.100)
	4	4.854	4.219	3.904	3.347	2.856	2.259
		(.0010)	(.0049)	(.010)	(.025)	(.050)	(.100)
	5	5.265	4.573	4.216	3.615	3.080	2.435
		(.0010)	(.0050)	(.010)	(.025)	(.050)	(.100)
	6	5.594	4.860	4.469	3.831	3.265	2.578
		(.0010)	(.0050)	(.010)	(.025)	(.050)	(.100)
	7	5.853	5.062	4.655	3.993	3.393	2.678
		(.0010)	(.0050)	(.010)	(.025)	(.050)	(.100)
	8	6.010	5.200	4.773	4.100	3.486	2.748
		(.0010)	(.0050)	(.010)	(.025)	(.050)	(.100)
	9	6.122	5.283	4.849	4.166	3.542	2.795
		(.0010)	(.0050)	(.010)	(.025)	(.050)	(.100)
	10	6.144	5.321	4.876	4.186	3.557	2.806
		(.0010)	(.0050)	(.010)	(.025)	(.050)	(.100)

The Terry–Hoeffding normal scores test. This test, like the Wilcoxon rank-sum test, is a two-sample test for identity of location of two populations. We are given two independent random samples $x_1, x_2, \ldots, x_m$ and $y_1, y_2, \ldots, y_n$ and suppose that $m \leqslant n$. The xs and ys are pooled and ranked in decreasing order of magnitude from 1 to N ($N = m+n$). To test the null hypothesis H_0 that the locations are identical against the *one-sided* alternative hypothesis H_1 that the xs tend to be *larger* than the ys we take as test statistic T the sum $\Sigma_{x_i} E(N, R_i)$ of the normal scores corresponding to the ranks R_i of the xs. We reject H_0 and accept H_1 if T exceeds the appropriate upper quantile.

The null distribution of T is symmetric about the mean 0. The table gives selected values of T with tail probabilities Q near 0.001, 0.005, 0.01, 0.025, 0.05 and 0.10 (exact probabilities in brackets).

If H_1 is the hypothesis that the xs tend to be *smaller* than the ys we take T to be $-\Sigma_{x_i} E(N, R_i)$.

If H_1 is the *two-sided* hypothesis that the locations of the X- and Y-populations are different, we take as test statistic $|T|$. The quantiles of $|T|$ are related to those of T; for $P > \frac{1}{2}$:

$$T_{[P]} = |T|_{[2P-1]} \quad (\text{e.g. } |T|_{[.90]} = T_{[.95]})$$

For $N > 20$

$$T \approx \sqrt{\left\{\frac{mn}{N} S(N)\right\} r(N-2)} \quad \text{where} \quad S(N) = \sum_{k=1}^{N} \{E(N, k)\}^2$$

and $r(N-2)$ is the product-moment correlation coefficient for $\nu = N-2$ degrees of freedom (p. 78). S is tabulated on p. 34. The approximation is fair provided that neither m/n nor n/N is small.

(A) UPPER TAIL PROBABILITIES Q OF THE STATISTIC S FOR THE FRIEDMAN TEST/KENDALL COEFFICIENT OF CONCORDANCE

k	n	S	Q	k	n	S	Q	k	n	S	Q	k	n	S	Q
3	2	8	.167	3	10	42	.135	4	2	18	.167	4	8	82	.106
3	3	14	.194			50	.092			20	.042			84	.099
		18	.028			56	.066	4	3	29	.148			100	.051
3	4	18	.125			62	.046			33	.075			102	.049
		24	.069			86	.012			35	.054			138	.011
		26	.042			96	.007			37	.033			140	.009
		32	.005−			98	.006			41	.017			154	.005+
3	5	24	.124			104	.003			45	.002			158	.004
		26	.093	3	11	54	.100+	4	4	40	.105	5	3	54	.117
		32	.039			56	.087			42	.094			56	.096
		38	.024			62	.062			50	.052			62	.056
		42	.008			72	.043			52	.036			64	.045
		50	.001			98	.011			62	.012			74	.015
3	6	26	.142			104	.006			64	.007			76	.008
		32	.072			114	.004			66	.006			78	.005
		38	.052	3	12	56	.108			68	.003			80	.004
		42	.029			62	.080	4	5	51	.107	5	4	74	.113
		50	.012			74	.051			53	.093			76	.095
		54	.008			78	.038			61	.055			86	.060
		56	.006			104	.011			65	.044			88	.049
		62	.002			114	.007			81	.012			110	.010+
3	7	32	.112			122	.005−			83	.009			112	.008
		38	.085	3	13	56	.129			89	.005+			118	.005+
		42	.051			62	.098			91	.003			120	.004
		50	.027			78	.050+	4	6	62	.108	5	5	94	.107
		56	.016			86	.037			64	.089			96	.094
		62	.008			114	.012			74	.056			110	.056
		72	.004			122	.009			76	.043			112	.049
3	8	38	.120			128	.006			100	.011			144	.010+
		42	.079			134	.004			102	.010−			146	.009
		50	.047	3	14	62	.117			110	.006			154	.006
		62	.018			72	.089			114	.004			156	.005
		72	.010−			78	.063	4	7	73	.101			158	.004
		74	.008			86	.049			75	.093				
		78	.005−			126	.010+			89	.052				
3	9	42	.107			128	.009			91	.041				
		50	.069			134	.007			117	.012				
		54	.057			146	.005−			121	.009				
		56	.048	3	15	72	.106			131	.005+				
		78	.010+			74	.096			133	.004				
		86	.006			86	.059								
		96	.004			96	.047								
						128	.011								
						134	.010−								
						146	.007								
						150	.005−								

Sources: M. Friedman, *Annals of Mathematical Statistics* 11 (1940), 86–92; *Handbook of Statistical Tables*. D. B. Owen, Table 14.1; *Nonparametric Statistical methods*, M. Hollander & D. A. Wolfe, Table A15. There are several discrepancies.

(B) UPPER QUANTILES $S_{[P]}$ FOR THE FRIEDMAN TEST/ KENDALL COEFFICIENT OF CONCORDANCE

n	$k=3$		$k=4$		$k=5$		$k=6$		$k=7$	
	$S_{[.95]}$	$S_{[.99]}$	$S_{[.95]}$	$S_{[.99]}$	$S_{[.95]}$	$S_{[.99]}$	$S_{[.95]}$	$S_{[.99]}$	$S_{[.95]}$	$S_{[.99]}$
3							103.9	122.8	157.3	185.6
4							143.3	176.2	217.0	265.0
5					112.3	142.8	182.4	229.4	276.2	343.8
6					136.1	176.1	221.4	282.4	335.2	422.6
8			101.7	137.4	183.7	242.7	299.0	388.3	453.1	579.9
10			127.8	175.3	231.2	309.1	376.7	494.0	571.0	737.0
15	89.8	131.0	192.9	269.8	349.8	475.2	570.5	758.2	864.9	1129.5
20	119.7	177.0	258.0	364.2	468.5	641.2	764.4	1022.2	1158.7	1521.9
$\chi^2_{[P]}(k-1)$	5.991	9.210	7.815	11.345	9.488	13.277	11.071	15.086	12.592	16.812

The Friedman test. Suppose that k treatments are ranked 1 to k by each of n individuals (groups or blocks). The ranks are set out in an $n \times k$ table, with columns relating to treatments, and rows to individuals. Let R_j denote the sum of ranks in column j. The Friedman test statistic is

$$T = \frac{S}{nk(k+1)/12}$$

where

$$S = \sum_{j=1}^{k} (R_j - \bar{R})^2 = \sum_{j=1}^{k} R_j^2 - \tfrac{1}{4}n^2 k(k+1)^2$$

We reject the null hypothesis of no significant difference between the treatments if T exceeds the appropriate quantile. The test is sensitive to differences between the mean ranks of the columns. For $k = 2$ the test is equivalent to the sign test.

The tables relate to the null distribution of S. Table (A) gives selected upper tail probabilities Q near 0.10, 0.05, 0.01 and 0.005. Table (B), which supplements table (A), gives approximate upper quantiles $S_{[.95]}$ and $S_{[.99]}$; linear interpolation with respect to n is reliable.

The Kendall coefficient of concordance. We suppose that k objects (treatments etc.) are ranked 1 to k by each of n judges (tests etc.). The ranks are set out in an $n \times k$ table as in the Friedman test, and S is defined in the same way. The coefficient of concordance is:

$$W = \frac{S}{n^2(k^3 - k)/12}$$

W was designed as a measure of agreement between judges rather than as a test statistic. For perfect concordance $W = 1$.

For values of n and k not covered by the tables

$$\frac{(n-1)W}{1-W} \sim F(\nu_1, \nu_2) \quad \text{approximately}$$

where $\nu_1 = k - 1 - (2/n)$ and $\nu_2 = (n-1)\nu_1$. Equivalently:

$$S_{[P]} \approx \frac{n^2(k^3 - k)}{12} \times \frac{F_{[P]}}{F_{[P]} + n - 1}$$

This approximation, corrected for continuity, was used in constructing table (B).

For $k > 7, n < 6$ the F-approximation, without the correction for continuity, should be used. For other values of k and n not covered by the tables:

$$T = n(k-1)W \sim \chi^2(k-1) \quad \text{approximately}$$

QUANTILES $T_{[P]}$ OF THE NUMBER-OF-RUNS TEST STATISTIC ($m \neq n$)

m	P	2	4	6	8	10	12	14	16	18	20	P
						n						
2	.005		5	5	5	5	5	5	5	5	5	.995
	.01		5	5	5	5	5	5	5	5	5	.99
	.025		5	5	5	5	5	5	5	5	5	.975
	.05		5	5	5	5	5	5	5	5	5	.95
4	.005	2		9	9	9	9	9	9	9	9	.995
	.01	2		9	9	9	9	9	9	9	9	.99
	.025	2		8	9	9	9	9	9	9	9	.975
	.05	2		8	9	9	9	9	9	9	9	.95
6	.005	2	2		12	13	13	13	13	13	13	.995
	.01	2	3		12	13	13	13	13	13	13	.99
	.025	2	3		11	12	12	13	13	13	13	.975
	.05	2	4		11	11	12	12	13	13	13	.95
8	.005	2	3	4		15	16	16	17	17	17	.995
	.01	2	3	4		14	15	16	16	17	17	.99
	.025	2	4	4		14	15	15	16	16	16	.975
	.05	3	4	5		13	14	15	15	15	16	.95
10	.005	2	3	4	5		18	18	19	20	20	.995
	.01	2	3	4	5		17	18	19	19	19	.99
	.025	2	4	5	6		16	17	18	18	19	.975
	.05	3	4	6	7		16	16	17	18	18	.95
12	.005	2	3	4	5	6		20	21	22	22	.995
	.01	2	4	5	6	7		20	21	21	22	.99
	.025	3	4	5	7	8		19	20	20	21	.975
	.05	3	5	6	7	8		18	19	20	20	.95
14	.005	2	3	5	6	7	8		23	24	24	.995
	.01	2	4	5	6	7	8		22	23	24	.99
	.025	3	4	6	7	8	9		21	22	23	.975
	.05	3	5	6	8	9	10		20	21	22	.95
16	.005	2	4	5	6	7	8	9		25	26	.995
	.01	2	4	5	7	8	9	10		25	25	.99
	.025	3	5	6	7	9	10	11		24	24	.975
	.05	3	5	7	8	9	11	12		23	24	.95
18	.005	2	4	5	7	8	9	10	11		28	.995
	.01	2	4	6	7	8	9	10	11		27	.99
	.025	3	5	6	8	9	10	11	12		26	.975
	.05	3	5	7	9	10	11	12	13		25	.95
20	.005	2	4	5	7	8	9	10	11	12		.995
	.01	3	4	6	7	9	10	11	12	13		.99
	.025	3	5	7	8	10	11	12	13	14		.975
	.05	3	5	7	9	10	12	13	14	15		.95

Upper quantiles are placed above the diagonal, lower quantiles below it. Note that m and n are interchangeable.

The mean and variance of the T-distribution are:

$$\mu_T = \frac{2mn}{m+n} + 1 \qquad \sigma_T^2 = \frac{2mn(2mn-m-n)}{(m+n)^2(m+n-1)}$$

For $m = n$

$$\mu_T = n+1 \qquad \sigma_T^2 = \frac{n(n-1)}{2n-1}$$

and upper quantiles are given by:

$$T_{[P]} = 2(n+1) - T_{[1-P]}$$

If $m > 20$ or $n > 20$ the distribution is approximately normal with mean μ_T and variance σ_T^2.

LOWER QUANTILES $T_{[P]}$ OF THE NUMBER-OF-RUNS TEST STATISTIC ($m = n$)

P	3	4	5	6	7	8	9	10	n 11	12	13	14	15	16	17	18
.005	2	2	2	3	4	4	5	6	6	7	8	8	9	10	11	12
.01	2	2	3	3	4	5	5	6	7	8	8	9	10	11	11	12
.025	2	2	3	4	4	5	6	7	8	8	9	10	11	12	12	13
.05	2	3	4	4	5	6	7	7	8	9	10	11	12	12	13	14

P	19	20	21	22	23	24	25	26	n 27	28	29	30	31	32	33	34
.005	12	13	14	15	15	16	17	18	19	19	20	21	22	23	24	24
.01	13	14	15	15	16	17	18	19	20	20	21	22	23	24	25	25
.025	14	15	16	17	17	18	19	20	21	22	23	23	24	25	26	27
.05	15	16	17	18	18	19	20	21	22	23	24	25	26	26	27	28

P	35	36	37	38	39	40	41	42	n 43	44	45	46	47	48	49	50
.005	25	26	27	28	29	30	30	31	32	33	34	35	36	36	37	38
.01	26	27	28	29	30	31	32	32	33	34	35	36	37	38	39	39
.025	28	29	30	31	31	32	33	34	35	36	37	38	39	39	40	41
.05	29	30	31	32	33	34	35	36	36	37	38	39	40	41	42	43

Source: For the tables on pp. 76, 77, F. S. Swed & C. Eisenhart, *Annals of Mathematical Statistics* **14** (1943), 66–87, Table 2.

The one-sample runs test. A sequence of observations is made. The test is designed to test the null hypothesis H_0 that the observations constitute a random sample from some population. Suppose that the observations are of two types a, b (e.g. $+$, $-$), m of type a and n of type b, and let T be the total number of runs in the sequence. (A run is a sequence of one or more observations of the same type.) The tables give quantiles of the null distribution of T. If the hypothesis H_1 alternative to H_0 is that observations tend to follow observations of the same type, we reject H_0 in favour of H_1 if the number of runs is less than the appropriate lower quantile. If H_1 is that observations tend to alternate, we reject H_0 if T exceeds the appropriate upper quantile.

If H_1 is the two-sided hypothesis that the observations are non-random, we use a two-tail test.

Example. In a period of 100 days 69 were dry and 31 were wet. The number of runs of dry days or wet days was 26. Is the sequence significantly non-random at level 0.05? For $m = 69$ and $n = 31$ $\mu_T = 43.78$ and $\sigma_T = 4.249$. Using the normal approximation we find that $T_{[.025]} \approx 35.0$. Since $26 < 35.0$ we conclude that the sequence of dry and wet days is non-random at significance level 0.05.

Runs above and below the median. If we have a sequence (a time sequence perhaps) of observations $x_1, x_2, \ldots, x_n$, a simple check of randomness against the broad alternative of non-randomness is obtained by dividing the observations into classes according to the criterion $x \gtrless M$, where M is the median of the xs, and counting the number of runs T. Since the numbers of observations in the two classes are equal we use the table above.

The Wald–Wolfowitz two-sample test. This is used to test the hypothesis that the distributions (assumed continuous) from which two given independent random samples $x_1, x_2, \ldots, x_m$ and $y_1, y_2, \ldots, y_n$ have been drawn are identical. The $m+n$ observations are pooled and arranged in ascending order of magnitude. The number of runs of xs and of ys is counted, and tested for randomness as in the one-sample test. In the two-sample test only the lower quantiles are used.

Example. See Example 1 on p. 66.

UPPER QUANTILES $r_{[P]}$ OF THE NULL DISTRIBUTION OF THE PEARSON CORRELATION COEFFICIENT

P	.90	.95	.975	.99	.995	.999	P	.90	.95	.975	.99	.995	.999
Q	.10	.05	.025	.01	.005	.001	Q	.10	.05	.025	.01	.005	.001
$2Q$	.20	.10	.05	.02	.01	.002	$2Q$	.20	.10	.05	.02	.01	.002
ν							ν						
1	.9511	.9877	.9969	$.9^351$	$.9^388$	$.9^551$	21	.277	.352	.413	.482	.526	.610
2	.800	.9000	.9500	.9800	.9900	.9980	22	.271	.344	.404	.472	.515	.599
3	.687	.805	.878	.9343	.9587	.9859	23	.265	.337	.396	.462	.505	.588
4	.608	.729	.811	.882	.9172	.9633	24	.260	.330	.388	.453	.496	.578
5	.551	.669	.754	.833	.875	.9350	25	.255	.323	.381	.445	.487	.568
6	.507	.621	.707	.789	.834	.9049	26	.250	.317	.374	.437	.479	.559
7	.472	.582	.666	.750	.798	.875	27	.245	.311	.367	.430	.471	.550
8	.443	.549	.632	.715	.765	.847	28	.241	.306	.361	.423	.463	.541
9	.419	.521	.602	.685	.735	.820	29	.237	.301	.355	.416	.456	.533
10	.398	.497	.576	.658	.708	.795	30	.233	.296	.349	.409	.449	.526
11	.380	.476	.553	.634	.684	.772	35	.216	.275	.325	.381	.418	.492
12	.365	.458	.532	.612	.661	.750	40	.202	.257	.304	.358	.393	.463
13	.351	.441	.514	.592	.641	.730	45	.190	.243	.288	.338	.372	.439
14	.338	.426	.497	.574	.623	.711	50	.181	.231	.273	.322	.354	.419
15	.327	.412	.482	.558	.606	.694	60	.165	.211	.250	.295	.325	.385
16	.317	.400	.468	.543	.590	.678	70	.153	.195	.232	.274	.302	.358
17	.308	.389	.456	.529	.575	.662	80	.143	.183	.217	.257	.283	.336
18	.299	.378	.444	.516	.561	.648	90	.135	.173	.205	.242	.267	.318
19	.291	.369	.433	.503	.549	.635	100	.128	.164	.195	.230	.254	.303
20	.284	.360	.423	.492	.537	.622	120	.117	.150	.178	.210	.232	.277
							$z_{[P]}$	1.282	1.645	1.960	2.326	2.576	3.090

For a sample (x_i, y_i), $i = 1, 2, \ldots, n$, *Pearson's product-moment correlation coefficient* is defined by:

$$r = \frac{\Sigma_i (x_i - \bar{x})(y_i - \bar{y})}{\sqrt{\Sigma_i (x_i - \bar{x})^2}\sqrt{\Sigma_i (y_i - \bar{y})^2}}$$

This is known also as the *linear correlation coefficient*.

The table gives the null distribution of r for random samples of size n taken from a normal bivariate population under the null hypothesis H_0 that the population correlation coefficient ρ is zero. The number of degrees of freedom is $\nu = n-2$. For a partial correlation coefficient in a linear regression after s explanatory variables have been eliminated $\nu = n-s-2$; e.g. $s = 1$ for $r_{yx_1 \cdot x_2}$.

Linear interpolation with respect to ν is liable to an error of at most 2 units in the third decimal place. Lower quantiles are given by:

$$r_{[P]} = -r_{[1-P]}$$

For $\nu > 120$ the distribution is approximately normal with variance $1/(\nu+1)$; the distribution of

$$z = (\nu+1)^{\frac{1}{2}} r(\nu)$$

is approximately N(0, 1), and tail probabilities can be found by referring z to the table on p. 31. Quantiles are given by

$$r_{[P]}(\nu) \approx z_{[P]}(\nu+1)^{-\frac{1}{2}}$$

where $z_{[P]}$ is a quantile of N(0, 1).

The table can be used to test H_0 against the alternatives $\rho > 0$, or $\rho < 0$ (one-tail tests), or $\rho \neq 0$ (two-tail test).

$r(\nu)$ is related to $t(\nu)$:

$$r\sqrt{\left(\frac{\nu}{1-r^2}\right)} = \sqrt{\nu}\ \tan\sin^{-1}r = \sqrt{\nu}\ \sinh\tanh^{-1}r = t$$

Inversely:

$$r_{[P]} = \frac{t_{[P]}}{\sqrt{(t_{[P]}^2+\nu)}} = \sin\tan^{-1}(t_{[P]}/\sqrt{\nu}) = \tanh\sinh^{-1}(t_{[P]}/\sqrt{\nu})$$

THE Z-TRANSFORMATION

r	0	1	2	3	4	5	6	7	8	9
.0	0.000	0.010	0.020	0.030	0.040	0.050	0.060	0.070	0.080	0.090
.1	0.100	0.110	0.121	0.131	0.141	0.151	0.161	0.172	0.182	0.192
.2	0.203	0.213	0.224	0.234	0.245	0.255	0.266	0.277	0.288	0.299
.3	0.310	0.321	0.332	0.343	0.354	0.365	0.377	0.388	0.400	0.412
.4	0.424	0.436	0.448	0.460	0.472	0.485	0.497	0.510	0.523	0.536
.5	0.549	0.563	0.576	0.590	0.604	0.618	0.633	0.648	0.662	0.678
.6	0.693	0.709	0.725	0.741	0.758	0.775	0.793	0.811	0.829	0.848
.7	0.867	0.887	0.908	0.929	0.950	0.973	0.996	1.020	1.045	1.071
.8	1.099	1.127	1.157	1.188	1.221	1.256	1.293	1.333	1.376	1.422
.90	1.472	1.478	1.483	1.488	1.494	1.499	1.505	1.510	1.516	1.522
.91	1.528	1.533	1.539	1.545	1.551	1.557	1.564	1.570	1.576	1.583
.92	1.589	1.596	1.602	1.609	1.616	1.623	1.630	1.637	1.644	1.651
.93	1.658	1.666	1.673	1.681	1.689	1.697	1.705	1.713	1.721	1.730
.94	1.738	1.747	1.756	1.764	1.774	1.783	1.792	1.802	1.812	1.822
.95	1.832	1.842	1.853	1.863	1.874	1.886	1.897	1.909	1.921	1.933
.96	1.946	1.959	1.972	1.986	2.000	2.014	2.029	2.044	2.060	2.076
.97	2.092	2.110	2.127	2.146	2.165	2.185	2.205	2.227	2.249	2.273
.98	2.298	2.323	2.351	2.380	2.410	2.443	2.477	2.515	2.555	2.599
.990	2.647	2.652	2.657	2.662	2.667	2.672	2.678	2.683	2.689	2.694
.991	2.700	2.705	2.711	2.717	2.722	2.728	2.734	2.740	2.746	2.752
.992	2.759	2.765	2.771	2.778	2.784	2.791	2.798	2.805	2.812	2.819
.993	2.826	2.833	2.840	2.848	2.855	2.863	2.871	2.879	2.887	2.895
.994	2.903	2.911	2.920	2.929	2.938	2.947	2.956	2.965	2.975	2.985
.995	2.994	3.005	3.015	3.025	3.036	3.047	3.059	3.070	3.082	3.094
.996	3.106	3.119	3.132	3.145	3.159	3.173	3.188	3.203	3.218	3.234
.997	3.250	3.267	3.285	3.303	3.322	3.342	3.362	3.383	3.406	3.429
.998	3.453	3.479	3.506	3.535	3.565	3.597	3.632	3.669	3.709	3.753

The tabulated function is:

$$Z = \tanh^{-1}r = \tfrac{1}{2}\ln\frac{1+r}{1-r} \approx 1.1513\lg\frac{1+r}{1-r}$$

If r is the product-moment correlation coefficient of a sample of n pairs of observations taken from a bivariate population whose correlation coefficient is ρ, the distribution of Z is, for large n, approximately normal with mean $\tanh^{-1}\rho+\rho/2(\nu+1)$ and variance $1/(\nu-1)$, with $\nu = n-2$ or, for a partial correlation coefficient in a regression after s variables have been eliminated, $\nu = n-s-2$. This 'variance stabilising' transformation is useful in the determination of the significance of the difference between the correlation coefficients of two samples or between the correlation coefficient of a sample and a hypothesised coefficient.

LOWER QUANTILES $S_{[P]}$ OF THE SPEARMAN/ HOTELLING–PABST TEST STATISTIC

n	.001	.005	.01	.025	.05	.10	$\frac{1}{6}n(n^2-1)$
4	0	0	0	0	2	2	10
5	0	0	2	2	4	6	20
6	0	2	4	6	8	14	35
7	2	6	8	14	18	26	56
8	6	12	16	24	32	42	84
9	12	22	28	38	50	64	120
10	22	36	44	60	74	92	165
11	36	56	66	86	104	128	220
12	52	78	94	120	144	172	286
13	76	110	130	162	190	226	364
14	106	148	172	212	246	290	455
15	142	194	224	270	312	364	560
16	186	250	284	340	390	450	680
17	238	314	356	420	480	550	816
18	300	390	438	512	582	664	969
19	372	476	532	618	696	790	1140
20	454	574	638	738	826	934	1330
21	546	686	758	870	972	1092	1540
22	652	810	892	1020	1134	1270	1771
23	772	950	1042	1184	1312	1464	2024
24	904	1104	1208	1366	1510	1678	2300
25	1050	1274	1390	1566	1726	1912	2600
26	1212	1462	1590	1786	1960	2168	2925
27	1390	1666	1808	2024	2216	2444	3276
28	1586	1890	2046	2284	2494	2744	3654
29	1800	2134	2306	2564	2796	3068	4060
30	2032	2398	2584	2868	3120	3416	4495

Source: G. J. Glasser & R. F. Winter, *Biometrika* **48** (1961), 444–8, Table 2; *Practical Nonparametric Statistics*, W. J. Conover, Table 9.

Let (x_i, y_i), $i = 1, 2, \ldots, n$, be a random sample from a population of matched pairs. Denote by R_i the rank of x_i in a ranking of the xs from 1 to n, and by S_i the rank of y_i in a ranking of the ys. The *Spearman correlation coefficient* r_S is defined by:

$$r_S = 1 - \frac{S}{n(n^2-1)/6}$$

where

$$S = \sum_i (R_i - S_i)^2$$

If there are no ties, r_S is the product-moment correlation coefficient for the pairs (R_i, S_i).

The table gives lower quantiles of the distribution of S under the null hypothesis H_0 that X and Y are uncorrelated in the population. Upper quantiles are given by:

$$S_{[P]} = n(n^2-1)/3 - S_{[1-P]}$$

The mean variance of r_S are 0 and $1/(n-1)$.

For $n > 30$ the product-moment correlation coefficient r (with $\nu = n-2$) is a good approximation to r_S.

We use r_S as a measure of correlation when the data are measured on an ordinal scale. S is used as a test statistic. To test H_0 against the one-sided alternative H_1 that the population correlation is *positive*, we compute S and reject H_0 in favour of H_1 if S is less than the appropriate lower quantile. If H_1 is that the population correlation is *negative*, we reject H_0 in favour of H_1 if S exceeds the appropriate upper quantile. If H_1 is the two-sided alternative that X and Y are correlated, we use a two-tail test. A test using S is sometimes known as a *Hotelling–Pabst test*. For an example see p. 83.

The Daniels test. If a series of observations $y_1, y_2, \ldots, y_n$ is made at times $t_1, t_2, \ldots, t_n$ the Hotelling–Pabst test for correlation between pairs (t_i, y_i) can be used to test for trend (i.e. a tendency for Y to increase steadily or decrease steadily with time).

UPPER QUANTILES $T_{[P]}$ FOR THE KENDALL CORRELATION COEFFICIENT

n	P .90	.95	.975	.990	.995	$\frac{1}{2}n(n-1)$	n	P .90	.95	.975	.990	.995	$\frac{1}{2}n(n-1)$
							21	42	54	64	76	84	210
							22	45	59	69	81	89	231
							23	49	63	73	87	97	253
4	4	4	6	6	6	6	24	52	66	78	92	102	276
5	6	6	8	8	10	10	25	56	70	84	98	108	300
6	7	9	11	11	13	15	26	59	75	89	105	115	325
7	9	11	13	15	17	21	27	61	79	93	111	123	351
8	10	14	16	18	20	28	28	66	84	98	116	128	378
9	12	16	18	22	24	36	29	68	88	104	124	136	406
10	15	19	21	25	27	45	30	73	93	109	129	143	435
11	17	21	25	29	31	55	31	75	97	115	135	149	465
12	18	24	28	34	36	66	32	80	102	120	142	158	496
13	22	26	32	38	42	78	33	84	106	126	150	164	528
14	23	31	35	41	45	91	34	87	111	131	155	173	561
15	27	33	39	47	51	105	35	91	115	137	163	179	595
16	28	36	44	50	56	120	36	94	120	144	170	188	630
17	32	40	48	56	62	136	37	98	126	150	176	196	666
18	35	43	51	61	67	153	38	103	131	155	183	203	703
19	37	47	55	65	73	171	39	107	137	161	191	211	741
20	40	50	60	70	78	190	40	110	142	168	198	220	780
							$z_{[P]}$	1.282	1.695	1.960	2.326	2.576	

Source: Practical Nonparametric Statistics, W. J. Conover, Table 11.

Let (x_i, y_i), $i = 1, 2, \ldots, n$, be a random sample taken from a bivariate population. A pair (x_i, y_i) and (x_j, y_j) are said to be *concordant* if $x_i - x_j$ and $y_i - y_j$ have the same sign and *discordant* if they have opposite signs. Let N_c be the number of concordant pairs in the sample and N_d the number of discordant pairs. The *Kendall correlation coefficient* is:

$$\tau = \frac{T}{n(n-1)/2} \quad \text{where} \quad T = N_c - N_d$$

Tied pairs (i.e. pairs with equal X-values or equal Y-values) are not included in N_c or N_d, so that $N_c + N_d + \text{no. of ties} = \frac{1}{2}n(n-1)$. It is often easier to count N_c and N_d if the observations are arranged in order of increasing X-values, and the X- and Y-values are replaced by their ranks.

The table gives upper quantiles of the distribution of T under the null hypothesis H_0 that X and Y are uncorrelated in the population. The distribution is symmetric about its mean 0. Its variance is:

$$\sigma_T^2 = n(n-1)(2n+5)/18$$

Lower quantiles are given by:

$$T_{[P]} = -T_{[1-P]}$$

For $n > 40$ the distribution of T is approximately normal.

We use τ (just as we use r_S) as a measure of correlation when the data are measured on an ordinal scale. T is used as a test statistic. To test H_0 against the one-sided alternative H_1 that X and Y are *positively* correlated in the population, we compute T and reject H_0 in favour of H_1 if T exceeds the appropriate upper quantile. If H_1 is that X and Y are *negatively* correlated, we reject H_0 in favour of H_1 if T is less than the appropriate lower quantile. If H_1 is the two-sided hypothesis that X and Y are correlated, we use a two-tail test. An example is given on p. 83.

The Mann–Kendall test for trend. The Kendall test can be used as a test for trend in a time (or other) series in the same way as the Spearman/Hotelling–Pabst test is used in Daniels' test.

UPPER TAIL PROBABILITIES Q FOR THE NORMAL SCORES CORRELATION COEFFICIENT B

	$n = 4$		$n = 5$		$n = 6$	
B	Q	B	Q	B	Q
		3.1952	.0083	3.7259	.0097
		2.9502	.0250	3.3353	.0250
2.2941	.0417	2.7490	.0417	3.0170	.0486
1.9413	.0833	2.5039	.0583	3.0136	.0514
		2.2151	.0917	2.4982	.0972
		2.1733	.1250	2.4861	.1028

Source: S. Bhuchongkul, *Annals of Mathematical Statistics* **35** (1964), 138–49.

A random sample (x_i, y_i), $i = 1, 2, \ldots, n$, taken from a bivariate population, is given. We denote by R_i the rank of x_i in a ranking of the xs in increasing order of magnitude, and by S_i the rank of y_i in a similar ranking of the ys. As a measure of correlation we use the statistic

$$B = \sum_i E(n, R_i) E(n, S_i)$$

where $E(n, R_i)$ and $E(n, S_i)$ are normal scores (p. 32).

We use B also as a test statistic to test the null hypothesis that the xs and ys are uncorrelated in the population. The null distribution of B is symmetric about the mean 0. The table gives selected values of B with upper tail probabilities near 0.01, 0.05 and 0.10.

For $n > 6$

$$B \approx r(n-2)\, S(n) \quad \text{where} \quad S(n) = \sum_k E(n, k)^2$$

and $r(n-2)$ is the ordinary coefficient of correlation for $\nu = n-2$ degrees of freedom (p. 78).

THE QUADRANT-SUM TEST FOR ASSOCIATION

n	$N_{[.80]}$	$N_{[.90]}$	$N_{[.95]}$	$N_{[.99]}$	$N_{[.995]}$	$N_{[.999]}$
6	6	12	–	–	–	–
8	6	8	10	16	–	–
10	6	8	10	14	20	–
14	6	8	10	14	16	21
∞	6	8	10	13	14	18

Source: P. S. Olmstead & J. W. Tukey, *Annals of Mathematical Statistics* **18** (1947), 495–513.

The quadrant-sum test. A random sample of n observations (x_i, y_i) is given. They are plotted on a scatter diagram, which is then divided into quadrants by the median lines $x = M_X$ and $y = M_Y$. The quadrants are labelled $+$ or $-$ according to the sign of $(x - M_X)(y - M_Y)$. Points on the median lines are discarded. Now, starting with the point with the largest x-coordinate on the right-hand side of the diagram and moving to the left, we count the points before we have to cross the median line $y = M_Y$; to their number N_A we attach a sign according to the sign of the quadrant in which they lie. We repeat the process, starting from the left and moving right, and obtain the score $\pm N_B$. In a similar way we obtain scores $\pm N_C$ and $\pm N_D$ by starting from above and moving down, and starting from below and moving up. The test statistic is the quadrant sum:

$$N = |\pm N_A \pm N_B \pm N_C \pm N_D|$$

The tabulated quantiles are computed on the null hypothesis H_0 that X and Y are independent. If the relevant quantile is exceeded, H_0 is rejected in favour of the alternative hypothesis that X and Y are associated.

Example. The following pairs of observations are given:

i	1	2	3	4	5	6	7	8	9	10	11	12
x_i	22	30	40	22	14	27	38	26	46	15	15	31
y_i	37	21	41	22	18	21	36	27	41	27	20	38

The medians are $M_X = 26\frac{1}{2}$ and $M_Y = 27$. The quadrant counts are $N_A = 4$, $N_B = 2$ (note that points on the median lines are discarded and that the point (22, 22), which is tied with (22, 37), is not included in the count of points *before* we cross the median line), $N_C = 3$ and $N_D = 2$. The quadrant sum is:

$$N = |+4+2+3+2| = 11$$

After discarding two points, $n = 10$, so $N_{[.95]} = 10$ and $N_{[.99]} = 14$. Therefore there is significant association at level 0.05 but not at 0.01.

In practice there would be no advantage in using the crude

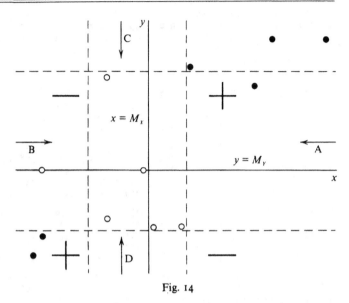

Fig. 14

Olmstead–Tukey test for as few as 12 observations, since the Hotelling–Pabst or Kendall tests would be as quick (pp. 80, 81).

The Hotelling–Pabst test. We restate the data using ranks R_i (for x), S_i (for y):

i	1	2	3	4	5	6	7	8	9	10	11	12
R_i	$4\frac{1}{2}$	8	11	$4\frac{1}{2}$	1	7	10	6	12	$2\frac{1}{2}$	$2\frac{1}{2}$	9
S_i	9	$3\frac{1}{2}$	$11\frac{1}{2}$	5	1	$3\frac{1}{2}$	8	$6\frac{1}{2}$	$11\frac{1}{2}$	$6\frac{1}{2}$	2	10
$\|R_i - S_i\|$	$4\frac{1}{2}$	$4\frac{1}{2}$	$\frac{1}{2}$	$\frac{1}{2}$	0	$3\frac{1}{2}$	2	$\frac{1}{2}$	$\frac{1}{2}$	4	$\frac{1}{2}$	1

$S = \Sigma(R_i - S_i)^2 = 75$. To test for association we use a two-sided test. Now, for $n = 12$, $S_{[.005]} = 78$. Since $75 < 78$ we conclude that there is significant association at level 0.01.

The Kendall test. We rearrange the data:

i	5	10	11	1	4	8	6	2	12	7	3	9
R_i	1	$2\frac{1}{2}$	$2\frac{1}{2}$	$4\frac{1}{2}$	$4\frac{1}{2}$	6	7	8	9	10	11	12
S_i	1	$6\frac{1}{2}$	2	9	5	$6\frac{1}{2}$	$3\frac{1}{2}$	$3\frac{1}{2}$	10	8	$11\frac{1}{2}$	$11\frac{1}{2}$

There are $N_c = 49$ concordant pairs and $N_d = 12$ discordant pairs, so that $T = N_c - N_d = 37$. To test for association we use a two-sided test. For $n = 12$, $T_{[.995]} = 36$. Since $37 > 36$, we conclude that there is significant association at level 0.01, in agreement with the Hotelling–Pabst test.

UPPER QUANTILES OF $\hat{R}^2$ IN MULTIPLE CORRELATION

P Q	k = 1			k = 2			k = 3			k = 4			k = 5		
	.90 .10	.95 .05	.99 .01	.90 .10	.95 .05	.99 .01	.90 .10	.95 .05	.99 .01	.90 .10	.95 .05	.99 .01	.90 .10	.95 .05	.99 .01
ν															
1	.976	.994		.990	.998		.994	.998		.996	.999		.997	.999	
2	.810	.903	.980	.900	.950	.990	.932	.966	.993	.949	.975	.995	.959	.980	.996
3	.649	.771	.919	.785	.864	.954	.844	.903	.967	.877	.924	.975	.898	.938	.979
4	.532	.658	.841	.684	.776	.900	.759	.832	.926	.804	.865	.941	.835	.887	.951
5	.448	.569	.765	.602	.698	.842	.685	.764	.879	.738	.806	.901	.775	.835	.916
6	.386	.499	.696	.536	.632	.785	.622	.704	.830	.680	.751	.859	.721	.785	.879
7	.339	.444	.636	.482	.575	.732	.568	.651	.784	.628	.702	.818	.673	.739	.842
8	.302	.399	.585	.438	.527	.684	.523	.604	.740	.584	.657	.778	.630	.697	.806
9	.272	.362	.540	.401	.486	.641	.484	.563	.700	.545	.618	.741	.592	.659	.771
10	.247	.332	.501	.369	.451	.602	.450	.527	.663	.510	.582	.706	.558	.624	.738
11	.227	.306	.467	.342	.420	.567	.420	.495	.629	.480	.550	.673	.527	.593	.707
12	.209	.283	.437	.319	.393	.536	.394	.466	.598	.453	.521	.643	.499	.564	.678
13	.194	.264	.411	.298	.369	.508	.371	.440	.570	.428	.494	.616	.474	.538	.652
14	.181	.247	.388	.280	.348	.482	.351	.417	.544	.406	.471	.590	.452	.514	.626
15	.170	.232	.367	.264	.329	.459	.332	.397	.520	.386	.449	.566	.431	.492	.603
16	.160	.219	.348	.250	.312	.438	.316	.378	.498	.368	.429	.544	.412	.471	.581
17	.151	.208	.331	.237	.297	.418	.301	.361	.478	.352	.411	.523	.395	.452	.560
18	.143	.197	.315	.226	.283	.401	.287	.345	.459	.337	.394	.504	.379	.435	.541
19	.136	.187	.301	.215	.270	.384	.275	.331	.442	.323	.379	.487	.364	.419	.523
20	.129	.179	.288	.206	.259	.369	.263	.317	.426	.310	.364	.470	.350	.404	.506
21	.124	.171	.276	.197	.248	.355	.253	.305	.410	.298	.351	.454	.338	.390	.490
22	.118	.164	.265	.189	.238	.342	.243	.294	.396	.288	.339	.440	.326	.377	.475
23	.113	.157	.255	.181	.229	.330	.234	.283	.383	.277	.327	.426	.315	.365	.461
24	.109	.151	.246	.175	.221	.319	.225	.273	.371	.268	.316	.413	.305	.353	.448
30	.088	.122	.201	.142	.181	.264	.185	.226	.311	.222	.264	.349	.255	.297	.381
40	.066	.093	.155	.109	.139	.206	.143	.176	.244	.173	.207	.277	.200	.234	.305
48	.055	.078	.130	.091	.117	.175	.121	.149	.209	.147	.176	.238	.170	.201	.263
60	.044	.063	.106	.074	.095	.142	.098	.121	.171	.120	.144	.196	.140	.165	.218
80	.033	.047	.080	.056	.072	.109	.075	.093	.131	.092	.111	.151	.107	.127	.169
120	.022	.032	.054	.038	.049	.074	.051	.063	.090	.062	.075	.104	.073	.087	.117
240	.011	.016	.027	.019	.025	.038	.026	.032	.046	.032	.039	.054	.038	.045	.061
∞	0	0	0	0	0	0	0	0	0	0	0	0	0	0	0

If $\hat{y}_i$ is the estimate of y_i in a linear regression on k explanatory variables, based on a sample of n observations ($i = 1, 2, \ldots, n$), the residual sum of squares is:

$$\sum_i (\hat{y}_i - y_i)^2 = \sum_i (y_i - \bar{y})^2 (1 - \hat{R}^2)$$

Here $\hat{R}$ is the product-moment correlation coefficient of the pairs $(\hat{y}_i, y_i)$; also $\hat{R}^2$ is equal to $\Sigma(\hat{y}_i - \bar{y})^2/\Sigma(y_i - \bar{y})^2$. $\hat{R}^2$ is a measure of the goodness of fit of the regression.

The table gives upper quantiles of the null distributions of $\hat{R}^2$ under the null hypothesis H_0 that $R = 0$ in the population. There are $\nu = n - k - 1$ degrees of freedom. Interpolation with respect to ν should be linear in $240/\nu$; errors will not exceed 2 units in the third decimal place. When $k = 1$, $\hat{R}$ is the ordinary correlation coefficient (but note that the upper tail of the $\hat{R}^2$-distribution corresponds to both tails of the r-distribution; e.g. $\hat{R}^2_{[.90]} = r^2_{[.95]}$).

If $\hat{R}^2$ is greater than the appropriate quantile, i.e. $\hat{R}^2$ is significantly different from 0, the regression model is useful.

The $\hat{R}^2$-distributions are related to F-distributions (p. 40):

$$\frac{\nu \hat{R}^2}{k(1 - \hat{R}^2)} = F(k, \nu) \qquad \hat{R}^2_{[P]} = \frac{k F_{[P]}}{k F_{[P]} + \nu}$$

LOWER QUANTILES OF THE VON NEUMANN RATIO v

n	P .001	.01	.05	n	P .001	.01	.05
5	0.416	0.538	0.820	35	1.030	1.249	1.459
6	0.363	0.561	0.890	36	1.042	1.258	1.466
7	0.370	0.614	0.936	37	1.053	1.267	1.473
8	0.404	0.663	0.982	38	1.064	1.276	1.479
9	0.442	0.709	1.024	39	1.075	1.285	1.486
10	0.482	0.752	1.062	40	1.085	1.293	1.492
11	0.520	0.791	1.097	41	1.095	1.302	1.498
12	0.556	0.828	1.128	42	1.105	1.310	1.504
13	0.590	0.862	1.156	43	1.114	1.317	1.510
14	0.622	0.893	1.182	44	1.123	1.325	1.515
15	0.653	0.922	1.205	45	1.132	1.332	1.521
16	0.683	0.949	1.227	46	1.140	1.339	1.526
17	0.710	0.974	1.247	47	1.148	1.345	1.530
18	0.737	0.998	1.266	48	1.156	1.351	1.535
19	0.762	1.020	1.283	49	1.164	1.357	1.539
20	0.785	1.041	1.300	50	1.171	1.363	1.544
21	0.807	1.060	1.315	51	1.177	1.368	1.548
22	0.828	1.078	1.329	52	1.184	1.374	1.552
23	0.848	1.096	1.343	53	1.191	1.379	1.556
24	0.867	1.112	1.355	54	1.198	1.385	1.560
25	0.885	1.128	1.367	55	1.204	1.390	1.563
26	0.902	1.143	1.378	56	1.210	1.395	1.567
27	0.918	1.157	1.389	57	1.217	1.400	1.571
28	0.934	1.170	1.399	58	1.223	1.405	1.574
29	0.950	1.183	1.409	59	1.229	1.410	1.578
30	0.965	1.195	1.418	60	1.235	1.414	1.581
31	0.979	1.207	1.427				
32	0.993	1.218	1.435				
33	1.006	1.228	1.443				
34	1.018	1.239	1.451				

Source: B. I. Hart, *Annals of Mathematical Statistics* **13** (1942), 207–14 and 445–7.

The von Neumann ratio. The statistic v is used to test for the presence of serial correlation in a series of observations $y_1, y_2, \ldots, y_n$:

$$v = \sum_{i=2}^{n} (y_i - y_{i-1})^2 \Big/ \sum_{i=1}^{n} (y_i - \bar{y})^2$$

For large n the ratio is nearly $2(1 - r_1)$ where r_1 is the coefficient of autocorrelation, i.e. correlation between pairs (y_i, y_{i-1}). The table gives lower quantiles of the null distribution of v under the null hypothesis H_0 that the observations are independent and normally distributed. H_0 can be rejected in favour of the alternative hypothesis H_1 that there is *positive* autocorrelation if v is less than the appropriate lower quantile. If the alternative is that there is *negative* autocorrelation we work with $4 - v$ instead of v.

The Durbin–Watson statistic. The definition of d is similar to that of v:

$$d = \sum_{i=2}^{n} (\hat{u}_i - \hat{u}_{i-1})^2 \Big/ \sum_{i=1}^{n} \hat{u}_i^2$$

where the $\hat{u}$s are the estimated residuals in a linear regression of Y on k explanatory variables. It is difficult to tabulate the quantiles of the null distribution of d under the null hypothesis H_0 of zero autocorrelation; upper bounds d_U and lower bounds d_L of lower quantiles are tabulated instead. H_0 is accepted against the alternative H_1 of *positive* autocorrelation if d is greater than the appropriate d_U and rejected in favour of H_1 if d is less than d_L. To test H_0 against the alternative of *negative* autocorrelation we work with $4 - d$ instead of d.

THE DURBIN–WATSON d-STATISTIC

n	P	$k=1$		$k=2$		$k=3$		$k=4$		$k=5$	
		d_L	d_U	d_L	d_U	d_L	d_U	d_L	d_U	d_L	d_U
15	.05	1.08	1.36	0.95	1.54	0.82	1.75	0.69	1.97	0.56	2.21
	.025	0.95	1.23	0.83	1.40	0.71	1.61	0.59	1.84	0.48	2.09
	.01	0.81	1.07	0.70	1.25	0.59	1.46	0.49	1.70	0.39	1.96
16	.05	1.10	1.37	0.98	1.54	0.86	1.73	0.74	1.93	0.62	2.15
	.025	0.98	1.24	0.86	1.40	0.75	1.59	0.64	1.80	0.53	2.03
	.01	0.84	1.09	0.74	1.25	0.63	1.44	0.53	1.66	0.44	1.90
17	.05	1.13	1.38	1.02	1.54	0.90	1.71	0.78	1.90	0.67	2.10
	.025	1.01	1.25	0.90	1.40	0.79	1.58	0.68	1.77	0.57	1.98
	.01	0.87	1.10	0.77	1.25	0.67	1.43	0.57	1.63	0.48	1.85
18	.05	1.16	1.39	1.05	1.53	0.93	1.69	0.82	1.87	0.71	2.06
	.025	1.03	1.26	0.93	1.40	0.82	1.56	0.72	1.74	0.62	1.93
	.01	0.90	1.12	0.80	1.26	0.71	1.42	0.61	1.60	0.52	1.80
19	.05	1.18	1.40	1.08	1.53	0.97	1.68	0.86	1.85	0.75	2.02
	.025	1.06	1.28	0.96	1.41	0.86	1.55	0.76	1.72	0.66	1.90
	.01	0.93	1.13	0.83	1.26	0.74	1.41	0.65	1.58	0.56	1.77
20	.05	1.20	1.41	1.10	1.54	1.00	1.68	0.90	1.83	0.79	1.99
	.025	1.08	1.28	0.99	1.41	0.89	1.55	0.79	1.70	0.70	1.87
	.01	0.95	1.15	0.86	1.27	0.77	1.41	0.68	1.57	0.60	1.74
21	.05	1.22	1.42	1.13	1.54	1.03	1.67	0.93	1.81	0.83	1.96
	.025	1.10	1.30	1.01	1.41	0.92	1.54	0.83	1.69	0.73	1.84
	.01	0.97	1.16	0.89	1.27	0.80	1.41	0.72	1.55	0.63	1.71
22	.05	1.24	1.43	1.15	1.54	1.05	1.66	0.96	1.80	0.86	1.94
	.025	1.12	1.31	1.04	1.42	0.95	1.54	0.86	1.68	0.77	1.82
	.01	1.00	1.17	0.91	1.28	0.83	1.40	0.75	1.54	0.66	1.69
23	.05	1.26	1.44	1.17	1.54	1.08	1.66	0.99	1.79	0.90	1.92
	.025	1.14	1.32	1.06	1.42	0.97	1.54	0.89	1.67	0.80	1.80
	.01	1.02	1.19	0.94	1.29	0.86	1.40	0.77	1.53	0.70	1.67
24	.05	1.27	1.45	1.19	1.55	1.10	1.66	1.01	1.78	0.93	1.90
	.025	1.16	1.33	1.08	1.43	1.00	1.54	0.91	1.66	0.83	1.79
	.01	1.04	1.20	0.96	1.30	0.88	1.41	0.80	1.53	0.72	1.66
25	.05	1.29	1.45	1.21	1.55	1.12	1.66	1.04	1.77	0.95	1.89
	.025	1.18	1.34	1.10	1.43	1.02	1.54	0.94	1.65	0.86	1.77
	.01	1.05	1.21	0.98	1.30	0.90	1.41	0.83	1.52	0.75	1.65
26	.05	1.30	1.46	1.22	1.55	1.14	1.65	1.06	1.76	0.98	1.88
	.025	1.19	1.35	1.12	1.44	1.04	1.54	0.96	1.65	0.88	1.76
	.01	1.07	1.22	1.00	1.31	0.93	1.41	0.85	1.52	0.78	1.64
27	.05	1.32	1.47	1.24	1.56	1.16	1.65	1.08	1.76	1.01	1.86
	.025	1.21	1.36	1.13	1.44	1.06	1.54	0.99	1.64	0.91	1.75
	.01	1.09	1.23	1.02	1.32	0.95	1.41	0.88	1.51	0.81	1.63
28	.05	1.33	1.48	1.26	1.56	1.18	1.65	1.10	1.75	1.03	1.85
	.025	1.22	1.37	1.15	1.45	1.08	1.54	1.01	1.64	0.93	1.74
	.01	1.10	1.24	1.04	1.32	0.97	1.41	0.90	1.51	0.83	1.62
29	.05	1.34	1.48	1.27	1.56	1.20	1.65	1.12	1.74	1.05	1.84
	.025	1.24	1.38	1.17	1.45	1.10	1.54	1.03	1.63	0.96	1.73
	.01	1.12	1.25	1.05	1.33	0.99	1.42	0.92	1.51	0.85	1.61
30	.05	1.35	1.49	1.28	1.57	1.21	1.65	1.14	1.74	1.07	1.83
	.025	1.25	1.38	1.18	1.46	1.12	1.54	1.05	1.63	0.98	1.73
	.01	1.13	1.26	1.07	1.34	1.01	1.42	0.94	1.51	0.88	1.61
31	.05	1.36	1.50	1.30	1.57	1.23	1.65	1.16	1.74	1.09	1.83
	.025	1.26	1.39	1.20	1.47	1.13	1.55	1.07	1.63	1.00	1.72
	.01	1.15	1.27	1.08	1.34	1.02	1.42	0.96	1.51	0.90	1.60

Source: For the tables on pp. 86–7, J. Durbin & G. S. Watson, *Biometrika* **38** (1951), 159–78, Tables 4, 5, 6.

THE DURBIN–WATSON d-STATISTIC

n	P	$k=1$ d_L	d_U	$k=2$ d_L	d_U	$k=3$ d_L	d_U	$k=4$ d_L	d_U	$k=5$ d_L	d_U
32	.05	1.37	1.50	1.31	1.57	1.24	1.65	1.18	1.73	1.11	1.82
	.025	1.27	1.40	1.21	1.47	1.15	1.55	1.08	1.63	1.02	1.71
	.01	1.16	1.28	1.10	1.35	1.04	1.43	0.98	1.51	0.92	1.60
33	.05	1.38	1.51	1.32	1.58	1.26	1.65	1.19	1.73	1.13	1.81
	.025	1.28	1.41	1.22	1.48	1.16	1.55	1.10	1.63	1.04	1.71
	.01	1.17	1.29	1.11	1.36	1.05	1.43	1.00	1.51	0.94	1.59
34	.05	1.39	1.51	1.33	1.58	1.27	1.65	1.21	1.73	1.15	1.81
	.025	1.29	1.41	1.24	1.48	1.17	1.55	1.12	1.63	1.06	1.70
	.01	1.18	1.30	1.13	1.36	1.07	1.43	1.01	1.51	0.95	1.59
35	.05	1.40	1.52	1.34	1.58	1.28	1.65	1.22	1.73	1.16	1.80
	.025	1.30	1.42	1.25	1.48	1.19	1.55	1.13	1.63	1.07	1.70
	.01	1.19	1.31	1.14	1.37	1.08	1.44	1.03	1.51	0.97	1.59
36	.05	1.41	1.52	1.35	1.59	1.29	1.65	1.24	1.73	1.18	1.80
	.025	1.31	1.43	1.26	1.49	1.20	1.56	1.15	1.63	1.09	1.70
	.01	1.21	1.32	1.15	1.38	1.10	1.44	1.04	1.51	0.99	1.59
37	.05	1.42	1.53	1.36	1.59	1.31	1.66	1.25	1.72	1.19	1.80
	.025	1.32	1.43	1.27	1.49	1.21	1.56	1.16	1.62	1.10	1.70
	.01	1.22	1.32	1.16	1.38	1.11	1.45	1.06	1.51	1.00	1.59
38	.05	1.43	1.54	1.37	1.59	1.32	1.66	1.26	1.72	1.21	1.79
	.025	1.33	1.44	1.28	1.50	1.23	1.56	1.17	1.62	1.12	1.70
	.01	1.23	1.33	1.18	1.39	1.12	1.45	1.07	1.52	1.02	1.58
39	.05	1.43	1.54	1.38	1.60	1.33	1.66	1.27	1.72	1.22	1.79
	.025	1.34	1.44	1.29	1.50	1.24	1.56	1.19	1.63	1.13	1.69
	.01	1.24	1.34	1.19	1.39	1.14	1.45	1.09	1.52	1.03	1.58
40	.05	1.44	1.54	1.39	1.60	1.34	1.66	1.29	1.72	1.23	1.79
	.025	1.35	1.45	1.30	1.51	1.25	1.57	1.20	1.63	1.15	1.69
	.01	1.25	1.34	1.20	1.40	1.15	1.46	1.10	1.52	1.05	1.58
45	.05	1.48	1.57	1.43	1.62	1.38	1.67	1.34	1.72	1.29	1.78
	.025	1.39	1.48	1.34	1.53	1.30	1.58	1.25	1.63	1.21	1.69
	.01	1.29	1.38	1.24	1.42	1.20	1.48	1.16	1.53	1.11	1.58
50	.05	1.50	1.59	1.46	1.63	1.42	1.67	1.38	1.72	1.34	1.77
	.025	1.42	1.50	1.38	1.54	1.34	1.59	1.30	1.64	1.26	1.69
	.01	1.32	1.40	1.28	1.45	1.24	1.49	1.20	1.54	1.16	1.59
55	.05	1.53	1.60	1.49	1.64	1.45	1.68	1.41	1.72	1.38	1.77
	.025	1.45	1.52	1.41	1.56	1.37	1.60	1.33	1.64	1.30	1.69
	.01	1.36	1.43	1.32	1.47	1.28	1.51	1.25	1.55	1.21	1.59
60	.05	1.55	1.62	1.51	1.65	1.48	1.69	1.44	1.73	1.41	1.77
	.025	1.47	1.54	1.44	1.57	1.40	1.61	1.37	1.65	1.33	1.69
	.01	1.38	1.45	1.35	1.48	1.32	1.52	1.28	1.56	1.25	1.60
70	.05	1.58	1.64	1.55	1.67	1.52	1.70	1.49	1.74	1.46	1.77
	.025	1.51	1.57	1.48	1.60	1.45	1.63	1.42	1.66	1.39	1.70
	.01	1.43	1.49	1.40	1.52	1.37	1.55	1.34	1.58	1.31	1.61
80	.05	1.61	1.66	1.59	1.69	1.56	1.72	1.53	1.74	1.51	1.77
	.025	1.54	1.59	1.52	1.62	1.49	1.65	1.47	1.67	1.44	1.70
	.01	1.47	1.52	1.44	1.54	1.42	1.57	1.39	1.60	1.36	1.62
90	.05	1.63	1.68	1.61	1.70	1.59	1.73	1.57	1.75	1.54	1.78
	.025	1.57	1.61	1.55	1.64	1.53	1.66	1.50	1.69	1.48	1.71
	.01	1.50	1.54	1.47	1.56	1.45	1.59	1.43	1.61	1.41	1.64
100	.05	1.65	1.69	1.63	1.72	1.61	1.74	1.59	1.76	1.57	1.78
	.025	1.59	1.63	1.57	1.65	1.55	1.67	1.53	1.70	1.51	1.72
	.01	1.52	1.56	1.50	1.58	1.48	1.60	1.46	1.63	1.44	1.65

The use of the Durbin–Watson statistic is described on p. 85.

-0.856	-0.063	0.787	-2.052	-1.192	-0.831	1.623	1.135	0.759	-0.189
-0.276	-1.110	0.752	-1.378	-0.583	0.360	0.365	1.587	0.621	1.344
0.379	-0.440	0.858	1.453	-1.356	0.503	-1.134	1.950	-1.816	-0.283
1.468	0.131	0.047	0.355	0.162	-1.491	-0.739	-1.182	-0.533	-0.497
-1.805	-0.772	1.286	-0.636	-1.312	-1.045	1.559	-0.871	-0.102	-0.123
2.285	0.554	0.418	-0.577	-1.489	-1.255	0.092	-0.597	-1.051	-0.980
-0.602	0.399	1.121	-1.026	0.087	1.018	-1.437	0.661	0.091	-0.637
0.229	-0.584	0.705	0.124	0.341	1.320	-0.824	-1.541	-0.163	2.329
1.382	-1.454	1.537	-1.299	0.363	-0.356	-0.025	0.294	2.194	-0.395
0.978	0.109	1.434	-1.094	-0.265	-0.857	-1.421	-1.773	0.570	-0.053
-0.678	-2.335	1.202	-1.697	0.547	-0.201	-0.373	-1.363	-0.081	0.958
-0.366	-1.084	-0.626	0.798	1.706	-1.160	-0.838	1.462	0.636	0.570
-1.074	-1.379	0.086	-0.331	-0.288	-0.309	-1.527	-0.408	0.183	0.856
-0.600	-0.096	0.696	0.446	1.417	-2.140	0.599	-0.157	1.485	1.387
0.918	1.163	-1.445	0.759	0.878	-1.781	-0.056	-2.141	-0.234	0.975
-0.791	-0.528	0.946	1.673	-0.680	-0.784	1.494	-0.086	-1.071	-1.196
0.598	-0.352	0.719	-0.341	0.056	-1.041	1.429	0.235	0.314	-1.693
0.567	-1.156	-0.125	-0.534	0.711	-0.511	0.187	-0.644	-1.090	-1.281
0.963	0.052	0.037	0.637	-1.335	0.055	0.010	-0.860	-0.621	0.713
0.489	-0.209	1.659	0.054	1.635	0.169	0.794	-1.550	1.845	-0.388
-1.627	-0.017	0.699	0.661	-0.073	0.188	1.183	-1.054	-1.615	-0.765
-1.096	1.215	0.320	0.738	1.865	-1.169	-0.667	-0.674	-0.062	1.378
-2.532	1.031	-0.799	1.665	-2.756	-0.151	-0.704	0.602	-0.672	1.264
0.024	-1.183	-0.927	-0.629	0.204	-0.825	0.496	2.543	0.262	-0.785
0.192	0.125	0.373	-0.931	-0.079	0.186	-0.306	0.621	-0.292	1.131
-1.324	-1.229	-0.648	-0.430	0.811	0.868	0.787	1.845	-0.374	-0.651
-0.726	-0.746	1.572	-1.420	1.509	-0.361	-0.310	-3.117	1.637	0.642
-1.618	1.082	-0.319	0.300	1.524	-0.418	-1.712	0.358	-1.032	0.537
1.695	0.843	2.049	0.388	-0.297	1.077	-0.462	0.655	0.940	-0.354
0.790	0.605	-3.077	1.009	-0.906	-1.004	0.693	-1.098	1.300	0.549
1.792	-0.895	-0.136	-1.765	1.077	0.418	-0.150	0.808	0.697	0.435
0.771	-0.741	-0.492	-0.770	-0.458	-0.021	1.385	-1.225	-0.066	-1.471
-1.438	0.423	-1.211	0.723	-0.731	0.883	-2.109	-2.455	-0.210	1.644
-0.294	1.266	-1.994	-0.730	0.545	0.397	1.069	-0.383	-0.097	-0.985
-1.966	0.909	0.400	0.685	-0.800	1.759	0.268	1.387	-0.414	1.615
-0.999	1.587	1.423	0.937	-0.943	0.090	1.185	-1.204	0.300	-1.354
0.581	0.481	-2.400	0.000	0.231	0.079	-2.842	-0.846	-0.508	-0.516
0.370	-1.452	-0.580	-1.462	-0.972	1.116	-0.994	0.374	-3.336	-0.058
0.834	-1.227	-0.709	-1.039	-0.014	-0.383	-0.512	-0.347	0.881	-0.638
-0.376	-0.813	0.660	-1.029	-0.137	0.371	0.376	0.968	1.338	-0.786
-1.621	0.815	-0.544	-0.376	-0.852	0.436	1.562	0.815	-1.048	0.188
0.163	-0.161	2.501	-0.265	-0.285	1.934	1.070	0.215	-0.876	0.073
1.786	-0.538	-0.437	0.324	0.105	-0.421	-0.410	-0.947	0.700	-1.006
2.140	1.218	-0.351	-0.068	0.254	0.448	-1.461	0.784	0.317	1.013
0.064	0.410	0.368	0.419	-0.982	1.371	0.100	-0.505	0.856	0.890
0.789	-0.131	1.330	0.506	-0.645	-1.414	2.426	1.389	-0.169	-0.194
-0.011	-0.372	-0.699	2.382	-1.395	-0.467	1.256	-0.585	-1.359	-1.804
-0.463	0.003	-1.470	1.493	0.960	0.364	-1.267	-0.007	0.616	0.624
-1.210	-0.669	0.009	1.284	-0.617	0.355	-0.589	-0.243	-0.015	-0.712
-1.157	0.481	0.560	1.287	1.129	-0.126	0.006	1.532	1.328	0.980

These tables give random values z of the random variable Z having the standard normal distribution: $Z \sim N(0, 1)$. The probability density function is:

$$\frac{1}{\sqrt{(2\pi)}} \exp\left(-\frac{1}{2}z^2\right) \qquad (-\infty < z < \infty)$$

1.556	0.119	−0.078	0.164	−0.455	0.077	−0.043	−0.299	0.249	−0.182
0.647	1.029	1.186	0.887	1.204	−0.657	0.644	−0.410	−0.652	−0.165
0.329	0.407	1.169	−2.072	1.661	0.891	0.233	−1.628	−0.762	−0.717
−1.188	1.171	−1.170	−0.291	0.863	−0.045	−0.205	0.574	−0.926	1.407
−0.917	−0.616	−1.589	1.184	0.266	0.559	−1.833	−0.572	−0.648	−1.090
0.414	0.469	−0.182	0.397	1.649	1.198	0.067	−1.526	−0.081	−0.192
0.107	−0.187	1.343	0.472	−0.112	1.182	0.548	2.748	0.249	0.154
−0.497	1.907	0.191	0.136	−0.475	0.458	0.183	−1.640	−0.058	1.278
0.501	0.083	−0.321	1.133	1.126	−0.299	1.299	1.617	1.581	2.455
−1.382	−0.738	1.225	1.564	−0.363	−0.548	1.070	0.390	−1.398	0.524
−0.590	0.699	−0.162	−0.011	1.049	−0.689	1.225	0.339	−0.539	−0.445
−1.125	1.111	−1.065	0.534	0.102	0.425	−1.026	0.695	−0.057	0.795
0.849	0.169	−0.351	0.584	2.177	0.009	−0.696	−0.426	−0.692	−1.638
−1.233	−0.585	0.306	0.773	1.304	−1.304	0.282	−1.705	0.187	−0.880
0.104	−0.468	0.185	0.498	−0.624	−0.322	−0.875	1.478	−0.691	−0.281
0.261	−1.883	−0.181	1.675	−0.324	−1.029	−0.185	0.004	−0.101	−1.187
−0.007	1.280	0.568	−1.270	1.405	1.731	2.072	1.686	0.728	−0.417
0.794	−0.111	0.040	−0.536	−0.976	2.192	1.609	−0.190	−0.279	−1.611
0.431	−2.300	−1.081	−1.370	2.943	0.653	−2.523	0.756	0.886	−0.983
−0.149	1.294	−0.580	0.482	−1.449	−1.067	1.996	−0.274	0.721	0.490
−0.216	−1.647	1.043	0.481	−0.011	−0.587	−0.916	−1.016	−1.040	−1.117
1.604	−0.851	−0.317	−0.686	−0.008	1.939	0.078	−0.465	0.533	0.652
−0.212	0.005	0.535	0.837	0.362	1.103	0.219	0.488	1.332	−0.200
0.007	−0.076	1.484	0.455	−0.207	−0.554	1.120	0.913	−0.681	1.751
−0.217	0.937	0.860	0.323	1.321	−0.492	−1.386	−0.003	−0.230	0.539
−0.649	0.300	−0.698	0.900	0.569	0.842	0.804	1.025	0.603	−1.546
−1.541	0.193	2.047	−0.552	1.190	−0.087	2.062	−2.173	−0.791	−0.520
0.274	−0.530	0.112	0.385	0.656	0.436	0.882	0.312	−2.265	−0.218
0.876	−1.498	−0.128	−0.387	−1.259	−0.856	−0.353	0.714	0.863	1.169
−0.859	−1.083	1.288	−0.078	−0.081	0.210	0.572	1.194	−1.118	−1.543
−0.015	−0.567	0.113	2.127	−0.719	3.256	−0.721	−0.663	−0.779	−0.930
−1.529	−0.231	1.223	0.300	−0.995	−0.651	0.505	0.138	−0.064	1.341
0.278	−0.058	−2.740	−0.296	−1.180	0.574	1.452	0.846	−0.243	−1.208
1.428	0.322	2.302	−0.852	0.782	−1.322	−0.092	−0.546	0.560	−1.430
0.770	−1.874	0.347	0.994	−0.485	−1.179	0.048	−1.324	1.061	0.449
−0.303	−0.629	0.764	0.013	−1.192	−0.475	−1.085	−0.880	1.738	−1.225
−0.263	−2.105	0.509	−0.645	1.362	0.504	−0.755	1.274	1.448	0.604
0.997	−1.187	−0.242	0.121	2.510	−1.935	0.350	0.073	0.458	−0.446
−0.063	−0.475	−1.802	−0.476	0.193	−1.199	0.339	0.364	−0.684	1.353
−0.168	1.904	−0.485	−0.032	−0.554	0.056	−0.710	−0.778	0.722	−0.024
0.366	−0.491	0.301	−0.008	−0.894	−0.945	0.384	−1.748	−1.118	0.394
0.436	−0.464	0.539	0.942	−0.458	0.445	−1.883	1.228	1.113	−0.218
0.597	−1.471	−0.434	0.705	−0.788	0.575	0.086	0.504	1.445	−0.513
−0.805	−0.624	1.344	0.649	−1.124	0.680	−0.986	1.845	−1.152	−0.393
1.681	−1.910	0.440	0.067	−1.502	−0.755	−0.989	−0.054	−2.320	0.474
−0.007	−0.459	1.940	0.220	−1.259	−1.729	0.137	−0.520	−0.412	2.847
0.209	−0.633	0.299	0.174	1.975	−0.271	0.119	−0.199	0.007	2.315
1.254	1.672	−1.186	−1.310	0.474	0.878	−0.725	−0.191	0.642	−1.212
−1.016	−0.697	0.017	−0.263	−0.047	−1.294	−0.339	2.257	−0.078	−0.049
−1.169	−0.355	1.086	−0.199	0.031	0.396	−0.143	1.572	0.276	0.027

2.988	0.423	−1.261	−1.893	0.187	−0.412	−0.228	0.002	−0.384	−1.032
0.760	0.995	−0.256	−0.505	0.750	−0.654	0.647	0.613	0.086	−0.118
−0.650	−0.927	−1.071	−0.796	1.130	−1.042	−0.181	−1.020	1.648	−1.327
−0.394	−0.452	0.893	1.410	1.133	0.319	0.537	−0.789	0.078	−0.062
−1.168	1.902	0.206	0.303	1.413	2.012	0.278	−0.566	−0.900	0.200
1.343	−0.377	−0.131	−0.585	0.053	0.137	−1.371	−0.175	−0.878	0.118
−0.733	−1.921	0.471	−1.394	−0.885	−0.523	0.553	0.344	−0.775	1.545
−0.172	−0.575	0.066	−0.310	1.795	−1.148	0.772	−1.063	0.818	0.302
1.457	0.862	1.677	−0.507	−1.691	−0.034	0.270	0.075	−0.554	1.420
−0.087	0.744	1.829	1.203	−0.436	−0.618	−0.200	−1.134	−1.352	−0.098
−0.092	1.043	−0.255	0.189	0.270	−1.034	−0.571	−0.336	−0.742	2.141
0.441	−0.379	−1.757	0.608	0.527	−0.338	−1.995	0.573	−0.034	−0.056
0.073	−0.250	0.531	−0.695	1.402	−0.462	−0.938	1.130	1.453	−0.106
0.637	0.276	−0.013	1.968	−0.205	0.486	0.727	1.416	0.963	1.349
−0.792	−1.778	1.284	−0.452	0.602	0.668	0.516	−0.210	0.040	−0.103
−1.223	1.561	−2.099	1.419	0.223	−0.482	1.098	0.513	0.418	−1.686
−0.407	1.587	0.335	−2.475	−0.284	1.567	−0.248	−0.759	1.792	−2.319
−0.462	−0.193	−0.012	−1.208	2.151	1.336	−1.968	−1.767	−0.374	0.783
1.457	0.883	1.001	−0.169	0.836	−1.236	1.632	−0.142	−0.222	0.340
−1.918	−1.246	−0.209	0.780	−0.330	−2.953	−0.447	−0.094	1.344	−0.196
−0.126	1.094	−1.206	−1.426	1.474	−1.080	0.000	0.764	1.476	−0.016
−0.306	−0.847	0.639	−0.262	−0.427	0.391	−1.298	−1.013	2.024	−0.539
0.477	1.595	−0.762	0.424	0.799	0.312	1.151	−1.095	1.199	−0.765
0.369	−0.709	1.283	−0.007	−1.440	−0.782	0.061	1.427	1.656	0.974
−0.579	0.606	−0.866	−0.715	−0.301	−0.180	0.188	0.668	−1.091	1.476
−0.418	−0.588	0.919	−0.083	1.084	0.944	0.253	−1.833	1.305	0.171
0.128	−0.834	0.009	0.742	0.539	−0.948	−1.055	−0.689	−0.338	1.091
−0.291	0.235	−0.971	−1.696	1.119	0.272	0.635	−0.792	−1.355	1.291
−1.024	1.212	−1.100	−0.348	1.741	0.035	1.268	0.192	0.729	−0.467
−0.378	1.026	0.093	0.468	−0.967	0.675	0.807	−2.109	−1.214	0.559
1.232	−0.815	0.608	1.429	−0.748	0.201	0.400	−1.230	−0.398	−0.674
1.793	−0.581	−1.076	0.512	−0.442	−1.488	−0.580	0.172	−0.891	0.311
0.766	0.310	−0.070	0.624	−0.389	1.035	−0.101	−0.926	0.816	−1.048
−0.606	−1.224	1.465	0.012	1.061	0.491	−1.023	1.948	0.866	−0.737
0.106	−2.715	0.363	0.343	−0.159	2.672	1.119	0.731	−1.012	−0.889
−0.060	0.444	1.596	−0.630	0.362	−0.306	1.163	−0.974	0.486	−0.373
2.081	1.161	−1.167	0.021	0.053	−0.094	0.381	−0.628	−2.581	−1.243
−1.727	−1.266	0.088	0.936	0.368	0.648	−0.799	1.115	−0.968	−2.588
0.091	1.364	1.677	0.644	1.505	0.440	−0.329	0.498	0.869	−0.965
−1.114	−0.239	−0.409	−0.334	−0.605	0.501	−1.921	−0.470	2.354	−0.660
0.189	−0.547	−1.758	−0.295	−0.279	−0.515	−1.053	0.553	−0.297	0.496
−0.065	−0.023	−0.267	−0.247	1.318	0.904	−0.712	−1.152	−0.543	0.176
−1.742	−0.599	0.430	−0.615	1.165	0.084	2.017	−1.207	2.614	1.490
0.732	0.188	2.343	0.526	−0.812	0.389	1.036	−0.023	0.229	−2.262
−1.490	0.014	0.167	1.422	0.015	0.069	0.133	0.897	−1.678	0.323
1.507	−0.571	−0.724	1.741	−0.152	−0.147	−0.158	−0.076	0.652	0.447
0.513	0.168	−0.076	−0.171	0.428	0.205	−0.865	0.107	1.023	0.077
−0.834	−1.121	1.441	0.492	0.559	1.724	−1.659	0.245	1.354	−0.041
0.258	1.880	−0.536	1.246	−0.188	−0.746	1.097	0.258	1.547	1.238
−0.818	0.273	0.159	−0.765	0.526	1.281	1.154	−0.687	−0.793	0.795

Random values of the random variable X having the normal distribution $N(\mu, \sigma^2)$ with probability density function

$$\frac{1}{\sqrt{(2\pi\sigma^2)}}\exp\left\{-\frac{1}{2}\left(\frac{x-\mu}{\sigma}\right)^2\right\} \qquad (-\infty < x < \infty)$$

can be derived from the tabulated values z: take $x = \mu + \sigma z$.

−1.752	−0.329	−1.256	0.318	1.531	0.349	−0.958	−0.059	0.415	−1.084
−0.291	0.085	1.701	−1.087	−0.443	−0.292	0.248	−0.539	−1.382	0.318
−0.933	0.130	0.634	0.899	1.409	−0.883	−0.095	0.229	0.129	0.367
−0.450	−0.244	0.072	1.028	1.730	−0.056	−1.488	−0.078	−2.361	−0.992
0.512	−0.882	0.490	−1.304	−0.266	0.757	−0.361	0.194	−1.078	0.529
−0.702	0.472	0.429	−0.664	−0.592	1.443	−1.515	−1.209	−1.043	0.278
0.284	0.039	−0.518	1.351	1.473	0.889	0.300	0.339	−0.206	1.392
−0.509	1.420	−0.782	−0.429	−1.266	0.627	−1.165	0.819	−0.261	0.409
−1.776	−1.033	1.977	0.014	0.702	−0.435	−0.816	1.131	0.656	0.061
−0.044	1.807	0.342	−2.510	1.071	−1.220	−0.060	−0.764	0.079	−0.964
0.263	−0.578	1.612	−0.148	−0.383	−1.007	−0.414	0.638	−0.186	0.507
0.986	0.439	−0.192	−0.132	0.167	0.883	−0.400	−1.440	−0.385	−1.414
−0.441	−0.852	−1.446	−0.605	−0.348	1.018	0.963	−0.004	2.504	−0.847
−0.866	0.489	0.097	0.379	0.192	−0.842	0.065	1.420	0.426	−1.191
−1.215	0.675	1.621	0.394	−1.447	2.199	−0.321	−0.540	−0.037	0.185
−0.475	−1.210	0.183	0.526	0.495	1.297	−1.613	1.241	−1.016	−0.090
1.200	0.131	2.502	0.344	−1.060	−0.909	−1.695	−0.666	−0.838	−0.866
−0.498	−1.202	−0.057	−1.354	−1.441	−1.590	0.987	0.441	0.637	−1.116
−0.743	0.894	−0.028	1.119	−0.598	0.279	2.241	0.830	0.267	−0.156
0.799	−0.780	−0.954	0.705	−0.361	−0.734	1.365	1.297	−0.142	−1.387
−0.206	−0.195	1.017	−1.167	−0.079	−0.452	0.058	−1.068	−0.394	−0.406
−0.092	−0.927	−0.439	0.256	0.503	0.338	1.511	−0.465	−0.118	−0.454
−1.222	−1.582	1.786	−0.517	−1.080	−0.409	−0.474	−1.890	0.247	0.575
0.068	0.075	−1.383	−0.084	0.159	1.276	1.141	0.186	−0.973	−0.266
0.183	1.600	−0.335	1.553	0.889	0.896	−0.035	0.461	0.486	1.246
−0.811	−2.904	0.618	0.588	0.533	0.803	−0.696	0.690	0.820	0.557
−1.010	1.149	1.033	0.336	1.306	0.835	1.523	0.296	−0.426	0.004
1.453	1.210	−0.043	0.220	−0.256	−1.161	−2.030	−0.046	0.243	1.082
0.759	−0.838	−0.877	−0.177	1.183	−0.218	−3.154	−0.963	−0.822	−1.114
0.287	0.278	−0.454	0.897	−0.122	0.013	0.346	0.921	0.238	−0.586
−0.669	0.035	−2.077	1.077	0.525	−0.154	−1.036	0.015	−0.220	0.882
0.392	0.106	−1.430	−0.204	−0.326	0.825	−0.432	−0.094	−1.566	0.679
−0.337	0.199	−0.160	0.625	−0.891	−1.464	−0.318	1.297	0.932	−0.032
0.369	−1.990	−1.190	0.666	−1.614	0.082	0.922	−0.139	−0.833	0.091
−1.694	0.710	−0.655	−0.546	1.654	0.134	0.466	0.033	−0.039	0.838
0.985	0.340	0.276	0.911	−0.170	−0.551	1.000	−0.838	0.275	−0.304
−1.063	−0.594	−1.526	−0.787	0.873	−0.405	−1.324	0.162	−0.163	−2.716
0.033	−1.527	1.422	0.308	0.845	−0.151	0.741	0.064	1.212	0.823
0.597	0.362	−3.760	1.159	0.874	−0.794	−0.915	1.215	1.627	−1.248
−1.601	−0.570	0.133	−0.660	1.485	0.682	−0.898	0.686	0.658	0.346
−0.266	−1.309	0.597	0.989	0.934	1.079	−0.656	−0.999	−0.036	−0.537
0.901	1.531	−0.889	−1.019	0.084	1.531	−0.144	−1.920	0.678	−0.402
−1.433	−1.008	−0.990	0.090	0.940	0.207	−0.745	0.638	1.469	1.214
1.327	0.763	−1.724	−0.709	−1.100	−1.346	−0.946	−0.157	0.522	−1.264
−0.248	0.788	−0.577	0.122	−0.536	0.293	1.207	−2.243	1.642	1.353
−0.401	−0.679	0.921	0.476	1.121	−0.864	0.128	−0.551	−0.872	1.511
0.344	−0.324	0.686	−1.487	−0.126	0.803	−0.961	0.183	−0.358	−0.184
0.441	−0.372	−1.336	0.062	1.506	−0.315	−0.112	−0.452	1.594	−0.264
0.824	0.040	−1.734	0.251	0.054	−0.379	1.298	−0.126	0.104	−0.529
1.385	1.320	−0.509	−0.381	−1.671	−0.524	−0.805	1.348	0.676	0.799

Source: For the tables on pp. 88–91, *A Million Random Digits*, The Rand Corporation.

RANDOM DIGITS

```
25 19 64 82 84    62 74 29 92 24    61 03 91 22 48    64 94 63 15 07    66 85 12 00 27
23 02 41 46 04    44 31 52 43 07    44 06 03 09 34    19 83 94 62 94    48 28 01 51 92
55 85 66 96 28    28 30 62 58 83    65 68 62 42 45    13 08 60 46 28    95 68 45 52 43
68 45 19 69 59    35 14 82 56 80    22 06 52 26 39    59 78 98 76 14    36 09 03 01 86
69 31 46 29 85    18 88 26 95 54    01 02 14 03 05    48 00 26 43 85    33 93 81 45 95

37 31 61 28 98    94 61 47 03 10    67 80 84 41 26    88 84 59 69 14    77 32 82 81 89
66 42 19 24 94    13 13 38 69 96    76 69 76 24 13    43 83 10 13 24    18 32 84 85 04
33 65 78 12 35    91 59 11 38 44    23 31 48 75 74    05 30 08 46 32    90 04 93 56 16
76 32 06 19 35    22 95 30 19 29    57 74 43 20 90    20 25 36 70 69    38 32 11 01 01
43 33 42 02 59    20 39 84 95 61    58 22 04 02 99    99 78 78 83 82    43 67 16 38 95

28 31 93 43 94    87 73 19 38 47    54 36 90 98 10    83 43 32 26 26    22 00 90 59 22
97 19 21 63 34    69 33 17 03 02    11 15 50 46 08    42 69 60 17 42    14 68 61 14 48
82 80 37 14 20    56 39 59 89 63    33 90 38 44 50    78 22 87 10 88    06 58 87 39 67
03 68 03 13 60    64 13 09 37 11    86 02 57 41 99    31 66 60 65 64    03 03 02 58 97
65 16 58 11 01    98 78 80 63 23    07 37 66 20 56    20 96 06 79 80    33 39 40 49 42

24 65 58 57 04    18 62 85 28 24    26 45 17 82 76    39 65 01 73 91    50 37 49 38 73
02 72 64 07 75    85 66 48 38 73    75 10 96 59 31    48 78 58 08 88    72 08 54 57 17
79 16 78 63 99    43 61 00 66 42    76 26 71 14 33    33 86 76 71 66    37 85 05 56 07
04 75 14 93 39    68 52 16 83 34    64 09 44 62 58    48 32 72 26 95    32 67 35 49 71
40 64 64 57 60    97 00 12 91 33    22 14 73 01 11    83 97 68 95 65    67 77 80 98 87

06 27 07 34 26    01 52 48 69 57    19 17 53 55 96    02 41 03 89 33    86 85 73 02 32
62 40 03 87 10    96 88 22 46 94    35 56 60 94 20    60 73 04 84 98    96 45 18 47 07
00 98 48 18 97    91 51 63 27 95    74 25 84 03 07    88 29 04 79 84    03 71 13 78 26
50 64 19 18 91    98 55 83 46 09    49 66 41 12 45    41 49 36 83 43    53 75 35 13 39
38 54 52 25 78    01 98 00 89 85    86 12 22 89 25    10 10 71 19 45    88 84 77 00 07

46 86 80 97 78    65 12 64 64 70    58 41 05 49 08    68 68 88 54 00    81 61 61 80 41
90 72 92 93 10    09 12 81 93 63    69 30 02 04 26    92 36 48 69 45    91 99 08 07 65
66 21 41 77 60    99 35 72 61 22    52 40 74 67 29    97 50 71 39 79    57 82 14 88 06
87 05 46 52 76    89 96 34 22 37    27 11 57 04 19    57 93 08 35 69    07 51 19 92 66
46 90 61 03 06    89 85 33 22 80    34 89 12 29 37    44 71 38 40 37    15 49 55 51 08

11 88 53 06 09    81 83 33 98 29    91 27 59 43 09    70 72 51 49 73    35 97 25 83 41
11 05 92 06 97    68 82 34 08 83    25 40 58 40 64    56 42 78 54 06    60 96 96 12 82
33 94 24 20 28    62 42 07 12 63    34 39 02 92 31    80 61 68 44 19    09 92 14 73 49
24 89 74 75 61    61 02 73 36 85    67 28 50 49 85    37 79 95 02 66    73 19 76 28 13
15 19 74 67 23    61 38 93 73 68    76 23 15 58 20    35 36 82 82 59    01 33 48 17 66

05 64 12 70 88    80 58 35 06 88    73 48 27 39 43    43 40 13 35 45    55 10 54 38 50
57 49 36 44 06    74 93 55 39 26    27 70 98 76 68    78 36 26 24 06    43 24 56 40 80
77 82 96 96 97    60 42 17 18 48    16 34 92 19 52    98 84 48 42 92    83 19 06 77 78
24 10 70 06 51    59 62 37 95 42    53 67 14 95 29    84 65 43 07 30    77 54 00 15 42
50 00 07 78 23    49 54 36 85 14    18 50 54 18 82    23 79 80 71 37    60 62 95 40 30

44 37 76 21 96    37 03 08 98 64    90 85 59 43 64    17 79 96 52 35    21 05 22 59 30
90 57 55 17 47    53 26 79 20 38    69 90 58 64 03    33 48 32 91 54    68 44 90 24 25
50 74 64 67 42    95 28 12 73 23    32 54 98 64 94    82 17 18 17 14    55 10 61 64 29
44 04 70 22 02    84 31 64 64 08    52 55 04 24 29    91 95 43 81 14    66 13 18 47 44
32 74 61 64 73    21 46 51 44 77    72 48 92 00 05    83 59 89 65 06    53 76 70 58 78

75 73 51 70 49    12 53 67 51 54    38 10 11 67 73    22 32 61 43 75    31 61 22 21 11
76 18 36 16 34    16 28 25 82 98    64 26 70 54 87    49 48 55 11 39    94 25 20 80 85
00 17 37 71 81    64 21 91 15 82    81 04 14 52 11    39 07 30 60 77    39 18 27 85 68
54 95 57 55 04    12 77 40 70 14    79 86 61 57 50    52 49 41 73 46    05 63 34 92 33
69 99 95 54 63    44 37 33 53 17    38 06 58 37 93    47 10 62 31 28    63 59 40 40 32
```

The digits are drawn randomly from an infinite population containing elements of ten types 0, 1, 2, . . . , 9 in equal proportions.

RANDOM DIGITS

28 89 65 87 08	13 50 63 04 23	25 47 57 91 13	52 62 24 19 94	91 67 48 57 10
30 29 43 65 42	78 66 28 55 80	47 46 41 90 08	55 98 78 10 70	49 92 05 12 07
95 74 62 60 53	51 57 32 22 27	12 72 72 27 77	44 67 32 23 13	67 95 07 76 30
01 85 54 96 72	66 86 65 64 60	56 59 75 36 75	46 44 33 63 71	54 50 06 44 75
10 91 46 96 86	19 83 52 47 53	65 00 51 93 51	30 80 05 19 29	56 23 27 19 03
05 33 18 08 51	51 78 57 26 17	34 87 96 23 95	89 99 93 39 79	11 28 94 15 52
04 43 13 37 00	79 68 96 26 60	70 39 83 66 56	62 03 55 86 57	77 55 33 62 02
05 85 40 25 24	73 52 93 70 50	48 21 47 74 63	17 27 27 51 26	35 96 29 00 45
84 90 90 65 77	63 99 25 69 02	09 04 03 35 78	19 79 95 07 21	02 84 48 51 97
28 55 53 09 48	86 28 30 02 35	71 30 32 06 47	93 74 21 86 33	49 90 21 69 74
89 83 40 69 80	97 96 47 59 97	56 33 24 87 36	17 18 16 90 46	75 27 28 52 13
73 20 96 05 68	93 41 69 96 07	97 50 81 79 59	42 37 13 81 83	92 42 85 04 31
10 89 07 76 21	40 24 74 36 42	40 33 04 46 24	35 63 02 31 61	34 59 43 36 96
91 50 27 78 37	06 06 16 25 98	17 78 80 36 85	26 41 77 63 37	71 63 94 94 33
03 45 44 66 88	97 81 26 03 89	39 46 67 21 17	98 10 39 33 15	61 63 00 25 92
89 41 58 91 63	65 99 59 97 84	90 14 79 61 55	56 16 88 87 60	32 15 99 67 43
13 43 00 97 26	16 91 21 32 41	60 22 66 72 17	31 85 33 69 07	68 49 20 43 29
71 71 00 51 72	62 03 89 26 32	35 27 99 18 25	78 12 03 09 70	50 93 19 35 56
19 28 15 00 41	92 27 73 40 38	37 11 05 75 16	98 81 99 37 29	92 20 32 39 67
56 38 30 92 30	45 51 94 69 04	00 84 14 36 37	95 66 39 01 09	21 68 40 95 79
39 27 52 89 11	00 81 06 28 48	12 08 05 75 26	03 35 63 05 77	13 81 20 67 58
73 13 28 58 01	05 06 42 24 07	60 60 29 99 93	72 93 78 04 36	25 76 01 54 03
81 60 84 51 57	12 68 46 55 89	60 09 71 87 89	70 81 10 95 91	83 79 68 20 66
05 62 98 07 85	07 79 26 69 61	67 85 72 37 41	85 79 76 48 23	61 58 87 08 05
62 97 16 29 18	52 16 16 23 56	62 95 80 97 63	32 25 34 03 36	48 84 60 37 65
31 13 63 21 08	16 01 92 58 21	48 79 74 73 72	08 64 80 91 38	07 28 66 61 59
97 38 35 34 19	89 84 05 34 47	88 09 31 54 88	97 96 86 01 69	46 13 95 65 96
32 11 78 33 82	51 99 98 44 39	12 75 10 60 36	80 66 39 94 97	42 36 31 16 59
81 99 13 37 05	08 12 60 39 23	61 73 84 89 18	26 02 04 37 95	96 18 69 06 30
45 74 00 03 05	69 99 47 26 52	48 06 30 00 18	03 30 28 55 59	66 10 71 44 05
11 84 13 69 01	88 91 28 79 50	71 42 14 96 55	98 59 96 01 36	88 77 90 45 59
14 66 12 87 22	59 45 27 08 51	85 64 23 85 41	64 72 08 59 44	67 98 36 65 56
40 25 67 87 82	84 27 17 30 37	48 69 49 02 58	98 02 50 58 11	95 39 06 35 63
44 48 97 49 43	65 45 53 41 07	14 83 46 74 11	76 66 63 60 08	90 54 33 65 84
41 94 54 06 57	48 28 01 83 84	09 11 21 91 73	97 28 44 74 06	22 30 95 69 72
07 12 15 58 84	93 18 31 83 45	54 52 62 29 91	53 58 54 66 05	47 19 63 92 75
64 27 90 43 52	18 26 32 96 83	50 58 45 27 57	14 96 39 64 85	73 87 96 76 23
80 71 86 41 03	45 62 63 40 88	35 69 34 10 94	32 22 52 04 74	69 63 21 83 41
27 06 08 09 92	26 22 59 28 27	38 58 22 14 79	24 32 12 38 42	33 56 90 92 57
54 68 97 20 54	33 26 74 03 30	74 22 19 13 48	30 28 01 92 49	58 61 52 27 03
02 92 65 68 99	05 53 15 26 70	04 69 22 64 07	04 73 25 74 82	78 35 22 21 88
83 52 57 78 62	98 61 70 48 22	68 50 64 55 75	42 70 32 09 60	58 70 61 43 97
82 82 76 31 33	85 13 41 38 10	16 47 61 43 77	83 27 19 70 41	34 78 77 60 25
38 61 34 09 49	04 41 66 09 76	20 50 73 40 95	24 77 95 73 20	47 42 80 61 03
01 01 11 88 38	03 10 16 82 24	39 58 20 12 39	82 77 02 18 88	33 11 49 15 16
21 66 14 38 28	54 08 18 07 04	92 17 63 36 75	33 14 11 11 78	97 30 53 62 38
32 29 30 69 59	68 50 33 31 47	15 64 88 75 27	04 51 41 61 96	86 62 93 66 71
04 59 21 65 47	39 90 89 86 77	46 86 86 88 86	50 09 13 24 91	54 80 67 78 66
38 64 50 07 36	56 50 45 94 25	48 28 48 30 51	60 73 73 03 87	68 47 37 10 84
48 33 50 83 53	59 77 64 59 90	58 92 62 50 18	93 09 45 89 06	13 26 98 86 29

Example. To obtain a random sequence from an infinite population containing elements of four types 0, 1, 2 and 3 in equal proportions, we select digits from the table sequentially (by rows or by columns say) rejecting digits 8 and 9 (8 being the greatest multiple of 4 that does not exceed 10) and taking the remaining digits *modulo* 4. Thus the top row gives 2 – – – 2 1 – 3 0 – ...

RANDOM DIGITS

10 27 53 96 23	71 50 54 36 23	54 31 04 82 98	04 14 12 15 09	26 78 25 47 47
28 41 50 61 88	64 85 27 20 18	83 36 36 05 56	39 71 65 09 62	94 76 62 11 89
34 21 42 57 02	59 19 18 97 48	80 30 03 30 98	05 24 67 70 07	84 97 50 87 46
61 81 77 23 23	82 82 11 54 08	53 28 70 58 96	44 07 39 55 43	42 34 43 39 28
61 15 18 13 54	16 86 20 26 88	90 74 80 55 09	14 53 90 51 17	52 01 63 01 59
91 76 21 64 64	44 91 13 32 97	75 31 62 66 54	84 80 32 75 77	56 08 25 70 29
00 97 79 08 06	37 30 28 59 85	53 56 68 53 40	01 74 39 59 73	30 19 99 85 48
36 46 18 34 94	75 20 80 27 77	78 91 69 16 00	08 43 18 73 68	67 69 61 34 25
88 98 99 60 50	65 95 79 42 94	93 62 40 89 96	43 56 47 71 66	46 76 29 67 02
04 37 59 87 21	05 02 03 24 17	47 97 81 56 51	92 34 86 01 82	55 51 33 12 91
63 62 06 34 41	94 21 78 55 09	72 76 45 16 94	29 95 81 83 83	79 88 01 97 30
78 47 23 53 90	34 41 92 45 71	09 23 70 70 07	12 38 92 79 43	14 85 11 47 23
87 68 62 15 43	53 14 36 59 25	54 47 33 70 15	59 24 48 40 35	50 03 42 99 36
47 60 92 10 77	88 59 53 11 52	66 25 69 07 04	48 68 64 71 06	61 65 70 22 12
56 88 87 59 41	65 28 04 67 53	95 79 88 37 31	50 41 06 94 76	81 83 17 16 33
02 57 45 86 67	73 43 07 34 48	44 26 87 93 29	77 09 61 67 84	06 69 44 77 75
31 54 14 13 17	48 62 11 90 60	68 12 93 64 28	46 24 79 16 76	14 60 25 51 01
28 50 16 43 36	28 97 85 58 99	67 22 52 76 23	24 70 36 54 54	59 28 61 71 96
63 29 62 66 50	02 63 45 52 38	67 63 47 54 75	83 24 78 43 20	92 63 13 47 48
45 65 58 26 51	76 96 59 38 72	86 57 45 71 46	44 67 76 14 55	44 88 01 62 12
39 65 36 63 70	77 45 85 50 51	74 13 39 35 22	30 53 36 02 95	49 34 88 73 61
73 71 98 16 04	29 18 94 51 23	76 51 94 84 86	79 93 96 38 63	08 58 25 58 94
72 20 56 20 11	72 65 71 08 86	79 57 95 13 91	97 48 72 66 48	09 71 17 24 89
75 17 26 99 76	89 37 20 70 01	77 31 61 95 46	26 97 05 73 51	53 33 18 72 87
37 48 60 82 29	81 30 15 39 14	48 38 75 93 29	06 87 37 78 48	45 56 00 84 47
68 08 02 80 72	83 71 46 30 49	89 17 95 88 29	02 39 56 03 46	97 74 06 56 17
14 23 98 61 67	70 52 85 01 50	01 84 02 78 43	10 62 98 19 41	18 83 99 47 99
49 08 96 21 44	25 27 99 41 28	07 41 08 34 66	19 42 74 39 91	41 96 53 78 72
78 37 06 08 43	63 61 62 42 29	39 68 95 10 96	09 24 23 00 62	56 12 80 73 16
37 21 34 17 68	68 96 83 23 56	32 84 60 15 31	44 73 67 34 77	91 15 79 74 58
14 29 09 34 04	87 83 07 55 07	76 58 30 83 64	87 29 25 58 84	86 50 60 00 25
58 43 28 06 36	49 52 83 51 14	47 56 91 29 34	05 87 31 06 95	12 45 57 09 09
10 43 67 29 70	80 62 80 03 42	10 80 21 38 84	90 56 35 03 09	43 12 74 49 14
44 38 88 39 54	86 97 37 44 22	00 95 01 31 76	17 16 29 56 63	38 78 94 49 81
90 69 59 19 51	85 39 52 85 13	07 28 37 07 61	11 16 36 27 03	78 86 72 04 95
41 47 10 25 62	97 05 31 03 61	20 26 36 31 62	68 69 86 95 44	84 95 48 46 45
91 94 14 63 19	75 89 11 47 11	31 56 34 19 09	79 57 92 36 59	14 93 87 81 40
80 06 54 18 66	09 18 94 06 19	98 40 07 17 81	22 45 44 84 11	24 62 20 42 31
67 72 77 63 48	84 08 31 55 58	24 33 45 77 58	80 45 67 93 82	75 70 16 08 24
59 40 24 13 27	79 26 88 86 30	01 31 60 10 39	53 58 47 70 93	85 81 56 39 38
05 90 35 89 95	01 61 16 96 94	50 78 13 69 36	37 68 53 37 31	71 26 35 03 71
44 43 80 69 98	46 68 05 14 82	90 78 50 05 62	77 79 13 57 44	59 60 10 39 66
61 81 31 96 82	00 57 25 60 59	46 72 60 18 77	55 66 12 62 11	08 99 55 64 57
42 88 07 10 05	24 98 65 63 21	47 21 61 88 32	27 80 30 21 60	10 92 35 36 12
77 94 30 05 39	28 10 99 00 27	12 73 73 99 12	49 99 57 94 82	96 88 57 17 91
78 83 19 76 16	94 11 68 84 26	23 54 20 86 85	23 86 66 99 07	36 37 34 92 09
87 76 59 61 81	43 63 64 61 61	65 76 36 95 90	18 48 27 45 68	27 23 65 30 72
91 43 05 96 47	55 78 99 95 24	37 55 85 78 78	01 48 41 19 10	35 19 54 07 73
84 97 77 72 73	09 62 06 65 72	87 12 49 03 60	41 15 20 76 27	50 47 02 29 16
87 41 60 76 83	44 88 96 07 80	83 05 83 38 96	73 70 66 81 90	30 56 10 48 59

Example. To obtain a random sequence from an infinite population containing elements of 17 types 0, 1, 2, ..., 16 in equal proportions, we select two-digit numbers from the table sequentially (by rows or by columns say), rejecting 85 to 99 (85 being the greatest multiple of 17 that does not exceed 100) and taking the remaining numbers *modulo* 17. Thus the top row gives 10 10 2 – 6 3 16 3 2 6 ... (To find a number *modulo* 17, we take the remainder after division by 17.)

RANDOM DIGITS

```
22 17 68 65 84    68 95 23 92 35    87 02 22 57 51    61 09 43 95 06    58 24 82 03 47
19 36 27 59 46    13 79 93 37 55    39 77 32 77 09    85 52 05 30 62    47 83 51 62 74
16 77 23 02 77    09 61 87 25 21    28 06 24 25 93    16 71 13 59 78    23 05 47 47 25
78 43 76 71 61    20 44 90 32 64    97 67 63 99 61    46 38 03 93 22    69 81 21 99 21
03 28 28 26 08    73 37 32 04 05    69 30 16 09 05    88 69 58 28 99    35 07 44 75 47

93 22 53 64 39    07 10 63 76 35    87 03 04 79 88    08 13 13 85 51    55 34 57 72 69
78 76 58 54 74    92 38 70 96 92    52 06 79 79 45    82 63 18 27 44    69 66 92 19 09
23 68 35 26 00    99 53 93 61 28    52 70 05 48 34    56 65 05 61 86    90 92 10 70 80
15 39 25 70 99    93 86 52 77 65    15 33 59 05 28    22 87 26 07 47    86 96 98 29 06
58 71 96 30 24    18 46 23 34 27    85 13 99 24 44    49 18 09 79 49    74 16 32 23 02

57 35 27 33 72    24 53 63 94 09    41 10 76 47 91    44 04 95 49 66    39 60 04 59 81
48 50 86 54 48    22 06 34 72 52    82 21 15 65 20    33 29 94 71 11    15 91 29 12 03
61 96 48 95 03    07 16 39 33 66    98 56 10 56 79    77 21 30 27 12    90 49 22 23 62
36 93 89 41 26    29 70 83 63 51    99 74 20 52 36    87 09 41 15 09    98 60 16 03 03
18 87 00 42 31    57 90 12 02 07    23 47 37 17 31    54 08 01 88 63    39 41 88 92 10

88 56 53 27 59    33 35 72 67 47    77 34 55 45 70    08 18 27 38 90    16 95 86 70 75
09 72 95 84 29    49 41 31 06 70    42 38 06 45 18    64 84 73 31 65    52 53 37 97 15
12 96 88 17 31    65 19 69 02 83    60 75 86 90 68    24 64 19 35 51    56 61 87 39 12
85 94 57 24 16    92 09 84 38 76    22 00 27 69 85    29 81 94 78 70    21 94 47 90 12
38 64 43 59 98    98 77 87 68 07    91 51 67 62 44    40 98 05 93 78    23 32 65 41 18

53 44 09 42 72    00 41 86 79 79    68 47 22 00 20    35 55 31 51 51    00 83 63 22 55
40 76 66 26 84    57 99 99 90 37    36 63 32 08 58    37 40 13 68 97    87 64 81 07 83
02 17 79 18 05    12 59 52 57 02    22 07 09 47 03    28 14 11 30 79    20 69 22 40 98
95 17 82 06 53    31 51 10 96 46    92 06 88 07 77    56 11 50 81 69    40 23 72 51 39
35 76 22 42 92    96 11 83 44 80    34 68 35 48 77    33 42 40 90 60    73 96 53 97 86

26 29 13 56 41    85 47 04 66 08    34 72 57 59 13    82 43 80 46 15    38 26 61 70 04
77 80 20 75 82    72 82 32 99 90    63 95 73 76 63    89 73 44 99 05    48 67 26 43 18
46 40 66 44 52    91 36 74 43 53    30 82 13 54 00    78 45 63 98 35    55 03 36 67 68
37 56 08 18 09    77 53 84 46 47    31 91 18 95 58    24 16 74 11 53    44 10 13 85 57
61 65 61 68 66    37 27 47 39 19    84 83 70 07 48    53 21 40 06 71    95 06 79 88 54

93 43 69 64 07    34 18 04 52 35    56 27 09 24 86    61 85 53 83 45    19 90 70 99 00
21 96 60 12 99    11 20 99 45 18    48 13 93 55 34    18 37 79 49 90    65 97 38 20 46
95 20 47 97 97    27 37 83 28 71    00 06 41 41 74    45 89 09 39 84    51 67 11 52 49
97 86 21 78 73    10 65 81 92 59    58 76 17 14 97    04 76 62 16 17    17 95 70 45 80
69 92 06 34 13    59 71 74 17 32    27 55 10 24 19    23 71 82 13 74    63 52 52 01 41

04 31 17 21 56    33 73 99 19 87    26 72 39 27 67    53 77 57 68 93    60 61 97 22 61
61 06 98 03 91    87 14 77 43 96    43 00 65 98 50    45 60 33 01 07    98 99 46 50 47
85 93 85 86 88    72 87 08 62 40    16 06 10 89 20    23 21 34 74 97    76 38 03 29 63
21 74 32 47 45    73 96 07 94 52    09 65 90 77 47    25 76 16 19 33    53 05 70 53 30
15 69 53 82 80    79 96 23 53 10    65 39 07 16 29    45 33 02 43 70    02 87 40 41 45

02 89 08 04 49    20 21 14 68 86    87 63 93 95 17    11 29 01 95 80    35 14 97 35 33
87 18 15 89 79    85 43 01 72 73    08 61 74 51 69    89 74 39 82 15    94 51 33 41 67
98 83 71 94 22    59 97 50 99 52    08 52 85 08 40    87 80 61 65 31    91 51 80 32 44
10 08 58 21 66    72 68 49 29 31    89 85 84 46 06    59 73 19 85 23    65 09 29 75 63
47 90 56 10 08    88 02 84 27 83    42 29 72 23 19    66 56 45 65 79    20 71 53 20 25

22 85 61 68 90    49 64 92 85 44    16 40 12 89 88    50 14 49 81 06    01 82 77 45 12
67 80 43 79 33    12 83 11 41 16    25 58 19 68 70    77 02 54 00 52    53 43 37 15 26
27 62 50 96 72    79 44 61 40 15    14 53 40 65 39    27 31 58 50 28    11 39 03 34 25
33 78 80 87 15    38 30 06 38 21    14 47 47 07 26    54 96 87 53 32    40 36 40 96 76
13 13 92 66 99    47 24 49 57 74    32 25 43 62 17    10 97 11 69 84    99 63 22 32 98
```

Source: For the tables on pp. 92–5, *Statistical Tables for Biological, Agricultural and Medical Research* (6th edn), Table 33, R. A. Fisher & F. Yates.

GUIDE TO TESTS

Object of test	(1) One sample $x_1 \ldots x_n$	(2) Sample of matched pairs (x_i, y_i)	(3) Two independent samples (see also col. 5)	(4) Sample of several matched observations	(5) Several independent samples
Goodness of fit Homogeneity	Normal scores p. 32 Kolmogorov (O) p. 67 χ^2-test (N) p. 47	McNemar (N) p. 25	Smirnov (O) p. 69 Wald–Wolfowitz (O) p. 77 Fisher (N) p. 49		χ^2-test (N) p. 48
Location (means, medians)	t-confidence intervals for mean p. 46 Signed-rank median test (I) p. 63 Sign test (O) p. 25	t-confidence intervals for difference between means p. 46 Signed-rank test (I) p. 63 Sign test (O) p. 25	t-confidence intervals for difference between means p. 46 Rank-sum test (O) p. 66 Normal scores (O) p. 72	Duncan multiple range p. 45 Friedman (O) p. 75 Cochran (N) p. 51	χ^2-median test (N) p. 48 Kruskal–Wallis test (O) p. 71
Dispersion	Estimate of σ from range p. 34 χ^2-confidence intervals for σ^2 p. 46		F-confidence intervals for variance ratio p. 47		Westenberg (O) p. 48
Independence (of criteria)		χ^2-test (N) p. 48 Fisher (N) p. 49			
Correlation Association	von Neumann (serial correlation) p. 85 Durbin–Watson (for a regression) p. 85	Pearson p. 78 Spearman (O) p. 80 Kendall (O) p. 81 Normal scores (O) p. 82 Cox–Stuart (O) p. 25 Quadrant-sum (O) p. 82 Fisher (O) p. 50 χ^2-test (O) p. 50		Multiple correlation p. 84 Kendall concordance p. 75	
Randomness Trend (in a sequence or time series)	Cox–Stuart (O) p. 25 Daniels (O) p. 80 Mann–Kendall (O) p. 81 Runs tests p. 77				

The letters N (*nominal*), O (*ordinal*), and I (*interval*) indicate the scale on which sample values must be recorded. (Values on a nominal scale can be placed in separate categories; values on an ordinal scale can be placed in order; the differences between values on an interval scale can be ordered.) For tests depending on the t- or F-distributions sample values should be taken from a normal distribution.